The Internet of Materials

T0133754

The Internet of Materials

Edited by
Christos Liaskos

CRC Press
Taylor & Francis Group
Boca Raton London New York

CRC Press is an imprint of the
Taylor & Francis Group, an **informa** business

First edition published 2021
by CRC Press
6000 Broken Sound Parkway NW, Suite 300, Boca Raton, FL 33487-2742

and by CRC Press
2 Park Square, Milton Park, Abingdon, Oxon, OX14 4RN

ISBN: 978-0-367-45738-9 (hbk)
ISBN: 978-1-003-04380-5 (ebk)

Typeset in CMR
by Nova Techset Private Limited, Bengaluru & Chennai, India

CONTENTS

Chapter 1 Preface 1

Chapter 2 Introduction 3

Chapter 3 Electromagnetic Specifications and Prototype
Designs of Software-Defined Surfaces 7

3.1   Electromagnetic Modeling of Metasurfaces 10
    3.1.1   Unit Cell, Polarizability, and Interaction
        Constant 10
    3.1.2   Impedance Boundary Condition 13
    3.1.3   Sheet Impedance 15
3.2   Metasurfaces and Their Functionalities Compared with
Other Sheet Materials and Phased Array Antennas 18
    3.2.1   Sheets of Usual Materials 19
    3.2.2   Antenna Arrays 20
    3.2.3   Metasurfaces 22
    3.2.4   Comparison of Possible Functionalities and
        Advantages of Metasurfaces 22
3.3   Tunable Metasurfaces: From Global Tuning to
Software-Defined Metasurfaces 25
    3.3.1   Global Tuning 25
        3.3.1.1   Electric Tuning 26
        3.3.1.2   Magnetic Tuning 27
        3.3.1.3   Tuning by Light 27
        3.3.1.4   Thermal Tuning 27
    3.3.2   Local Tuning 28
        3.3.2.1   Switch Diode 29
        3.3.2.2   Continuous Tuning Varactors 29
        3.3.2.3   Coding Metasurfaces 30
    3.3.3   Software-Defined Metasurfaces 31
3.4   Design Workflow of the Switch-Fabric Prototype 32
    3.4.1   Design Principles per Functionality 32
        3.4.1.1   Tunable Perfect Absorption 33
        3.4.1.2   Tunable Anomalous Reflection 34
    3.4.2   Proof-of-Concept Design and Its Performance ... 36
        3.4.2.1   Tunable Perfect Absorption 36
        3.4.2.2   Tunable Anomalous Reflection 37
    3.4.3   Practical Considerations and Restrictions 40

3.4.3.1 Electronic Package Considerations from the EM Aspect 42

3.4.3.2 Where Should We Position the Package Vertically? 43

3.4.3.3 Linking Electromagnetic and Package Designs 44

3.4.3.4 Basic Unit-Cell Parameters 49

3.4.3.5 Advanced Design Considerations 50

3.4.4 Overview of the Switch-Fabric Tunable Absorber Design 50

3.5 Electromagnetic Performance of the Switch-Fabric Design 51

3.5.1 Tunable Perfect Absorption 51

3.5.2 Anomalous Reflection 53

3.5.3 Polarization Conversion 55

3.5.4 Electromagnetic Characterization Procedures ... 57

3.5.4.1 Experimental Setup 57

3.5.4.2 Demonstration Procedure 59

3.5.4.3 Spatial Modulation of Load Configuration for Better Performance 61

3.6 Design of the Graphene-Based Prototype 62

3.6.1 Practical Considerations and Design Constraints 62

3.6.2 Graphene Combined with Metallic Patches 67

3.6.3 Frequency-Tunable Perfect Absorber 70

3.6.4 Switchable Absorber 72

3.6.5 All-Angle Perfect Absorber 73

3.7 Summary 75

Chapter 4 Designing the Internet-of-Materials Interaction Software 77

4.1 Software Design Considerations 78

4.2 The Internet-of-Materials Software Architecture 79

4.3 The Internet-of-Materials Application Programming Interface 84

4.3.1 General Use Case Diagram 84

4.3.2 The Database Diagram 88

4.3.2.1 Table "DoA" 92

4.3.2.2 Table "Polarities" 92

4.3.2.3 Table "SwitchStates" 93

4.3.2.4 Table "Physical Setup" 94

4.3.2.5 Table "Function Electromagnetic Profiles" 95

4.3.3 The Class Diagram 96

4.4   A Novel Software Class: The Electromagnetic Compiler   97
       4.4.1   A Qualitative View of the Compiling Process ...   97
       4.4.2   Metasurface Functions   100
       4.4.3   Formal Definition of a Metasurface Configuration   101
       4.4.4   Definition of Fitness Function   103
       4.4.5   Methods   104
4.5   Theoretical Foundations of the Electromagnetic
       Compiler   105
       4.5.1   Definitions   106
       4.5.2   Floquet (Unit-Cell) Analysis   107
       4.5.3   ABSORB Functionality   111
       4.5.4   REFLECT Functionality   113
       4.5.5   POLARIZE Functionality   113
       4.5.6   STEER Functionality   114
       4.5.7   SPLIT Functionality   117
       4.5.8   Far-Field Scattering/Radiation Pattern   118
       4.5.9   Formal Definition   118
       4.5.10 Semi-Analytical Calculation   119
       4.5.11 Polarization   121
       4.5.12 Scattered Power in a Lobe (Solid Angle Cone) ..   123
       4.5.13 Fitness Functions per Functionality   124
               4.5.13.1   ABSORB Functionality   124
               4.5.13.2   STEER Functionality   124
               4.5.13.3   REFLECT Functionality   127
               4.5.13.4   SPLIT Functionality   127
               4.5.13.5   POLARIZE Functionality   127
               4.5.13.6   FOCUS and COLLIMATE
                          Functionalities   128
               4.5.13.7   SCATTER Functionality   129
               4.5.13.8   ARBITRARY Functionality   130
       4.5.14 The Configuration Optimization Process   130
4.6   Software Aspects of the Electromagnetic Compiler   132
       4.6.1   General Use Cases   132
       4.6.2   Validating the Compilation Outcomes with
               Measurements   138
4.7   Conclusion   140

Chapter 5 Design of the HyperSurface Networking
          Aspects   141
5.1   Design Requirements of the HyperSurface Controller
       Network   143

5.2   HyperSurface Networking Components: The
       HyperSurface Network Controller 144
       5.2.1   HyperSurface Controller Communication 144
5.3   The HyperSurface Controller Network Topology 145
       5.3.1   HyperSurface Network Controller Addressing ... 146
       5.3.2   HyperSurface Network Controller Channel
                Mapping 146
5.4   HyperSurface Controller Network Communication
       Protocols 148
       5.4.1   Routing and Reporting Protocol 148
       5.4.2   Fault-Adaptive Routing 149
       5.4.3   Workload Characterization 152
5.5   Evaluation of the Controller Network Design and
       Performance via Simulations 153
       5.5.1   Custom-Built Simulations 153
       5.5.2   HyperSurface Controller Network Simulator 154
                5.5.2.1   The HyperSurface Controller
                           Network Simulation 154
       5.5.3   Formal Evaluation of the HSF-CN 156
       5.5.4   HyperSurface Emulator 159
5.6   The Controller-Gateway Communication Perspective ... 169
       5.6.1   Gateway Functionality 169
                5.6.1.1   Software/Firmware Design and
                           Development 177
                5.6.1.2   Tile Gateway Communication
                           Interface Firmware 179
                5.6.1.3   Error/Fault Detection 184
                5.6.1.4   Bluetooth Mesh Firmware 186
5.7   The HyperSurface within Control Loops 188
       5.7.1   System Model 189
       5.7.2   The Considered Model 190
       5.7.3   Control Algorithm 193
       5.7.4   Estimation Algorithm 193
       5.7.5   Performance Evaluation 195
5.8   Summary 196

Chapter 6 Internet of Things-Compliant Platforms for
           Inter-Networking Metamaterials 199
6.1   Overview 200
6.2   Hardware Actuation Approaches 201
       6.2.1   RF Switching Elements 201
                6.2.1.1   PIN Diodes 201

6.2.2   Controller to PIN Interface 202
        6.2.2.1   DAC 203
6.3   Controller Communication 205
    6.3.1   Controller to Controller Communication 206
        6.3.1.1   SPI 206
        6.3.1.2   I2C 207
        6.3.1.3   UART 207
        6.3.1.4   CAN 207
    6.3.2   Controller to Server Communication 208
        6.3.2.1   Bluetooth 208
        6.3.2.2   802.15.4 208
        6.3.2.3   Zigbee 210
        6.3.2.4   UWB 211
        6.3.2.5   LORA 211
6.4   Controller Hardware 212
    6.4.1   The ESP8266/ESP32 212
    6.4.2   Arduino 213
    6.4.3   Raspberry Pi 213
    6.4.4   BeagleBone 214
    6.4.5   Libelium Waspmote 214
    6.4.6   OpenMote 215
6.5   IoT Operating Systems 215
    6.5.1   TinyOS 216
    6.5.2   Contiki/Contiki-NG 217
    6.5.3   FreeRTOS 217
    6.5.4   Android Things 218
    6.5.5   OpenWrt 218
    6.5.6   Raspbian 219
    6.5.7   OpenWSN 219
6.6   IoT Broker 220
    6.6.1   MQTT Brokers 222
        6.6.1.1   Mosquitto 222
        6.6.1.2   RabbitMQ 222
        6.6.1.3   EMQ 222
        6.6.1.4   VerneMQ 223
6.7   Conclusions 223

Chapter 7 Interim: Drafting a Stack 225

Chapter 8 The Scaling Laws of HyperSurfaces 227
8.1   The HyperSurface Scalability versus Manufacturing
       Technologies 229

8.1.1   Scaling Model 231
       8.1.1.1   Dimensional Factors 231
       8.1.1.2   Programming Parameters 232
8.1.2   Methodology 232
       8.1.2.1   Unit Cell Model 234
       8.1.2.2   Metasurface Model 235
       8.1.2.3   Metasurface Coding 237
       8.1.2.4   Performance Metrics 238
       8.1.2.5   Validation 239
8.1.3   Performance Scalability 241
       8.1.3.1   Directivity 241
       8.1.3.2   Target Deviation 242
       8.1.3.3   Half Power Beam Width 243
       8.1.3.4   Side Lobe Level 244
8.1.4   The HyperSurface Energy Footprint, Cost, and
       Performance 244
       8.1.4.1   Cost and Power Models 245
       8.1.4.2   Application-Specific Figures of Merit 248
       8.1.4.3   Performance-Cost Analysis 249
8.2   The HyperSurface Data Traffic as a Scaling Concern ... 249
8.2.1   System Model 252
       8.2.1.1   Mobility Model 253
       8.2.1.2   Gateway Model 254
       8.2.1.3   Embedded Controller Network 254
8.2.2   Evaluation Methodology 255
       8.2.2.1   Relevant Inputs 256
       8.2.2.2   Traffic Analysis Metrics 256
       8.2.2.3   Walkthrough Example 257
8.2.3   Workload Characterization 257
       8.2.3.1   Spatio-Temporal Intensity 257
       8.2.3.2   Reconfiguration Delay 260
       8.2.3.3   Sensitivity Analysis 261
8.2.4   Indoor Mobility Scenario 263
8.3   Conclusions 265

**Chapter 9 Applications of the Internet of Materials:
Programmable Wireless Environments** 269
9.1   Deterministic Wireless Propagation Control as a
     Concept 270
9.2   Modeling, Simulating, and Configuring PWEs—A
     Ray-Routing Approach Based on Graph Theory 272

9.2.1 General Modeling and Properties of
HyperSurface Functions 272

9.2.2 Specialized Modeling of Function
Inputs/Outputs 274

9.2.3 Modeling Core HyperSurface Functions 277

9.2.4 A Graph Model for Simulating and Optimizing
Programmable Wireless Environments 283

9.2.5 Modeling Connectivity Objectives 286

9.2.5.1 Power Transfer Maximization 287

9.2.5.2 QoS Optimization 288

9.2.5.3 Eavesdropping Mitigation 289

9.2.5.4 Doppler Effect Mitigation 289

9.2.5.5 User Blocking 290

9.3 A K-Paths Approach for Multi-User Multi-Objective
Environment Configuration 290

9.4 Artificial Intelligence-Based Configuration of PWEs 294

9.4.1 Feed-Forward 296

9.4.2 Back-Propagation 296

9.5 The Novel PWE Potential in Communication Quality,
Cybersecurity, and Wireless Power Transfer 297

9.5.1 Multi-User Multi-Objective Showcase 300

9.5.2 Doppler Effect Mitigation Showcase 302

9.5.3 User Capacity and Stress Test 304

9.5.4 Evaluation of Neural Network-Based PWE
Heuristics 306

9.6 Conclusion 309

Chapter 10 Epilogue 319

Index 341

CHAPTER 1

Preface

Christos Liaskos, Ageliki Tsioliaridou, Sotiris Ioannidis

Foundation for Research and Technology Hellas, 71110, Heraklion, Crete, Greece

Over a long course of the last six years, a large group of scientists had the pleasure of exploring an exciting direction and perhaps a programmer's sci-fi dream: to "talk" to materials with software commands, and tune their physical properties accordingly. In the last three years, with the support of the European Union and the Future and Emerging Technologies program of Horizon 2020 and an excellent consortium, this concept was advanced significantly.

Writing a book about this new concept perhaps preceded any other of our joint intentions, and has been in our minds from the start. However, in this small foreword, a big thanks is due to the people that supported this effort from its very early stage, back in 2014:

Prof. Andreas Pitsillides from the University of Cyprus and Prof. Ian F. Akyildiz from Georgia Tech. We feel that without their continuous support and inspiration, this effort would have certainly remained fruitless, at least from our side co-authoring this short note.

Finally, while this is not a project-specific book, a big thanks to all VISORSURF project consortium members and extended collaborators in general [132], those who made it in contributing invited chapters to this book, those who could not make it due to other obligations, those who stayed until the end, and those who had to move on to their next professional step. Instead of listing and sorting them in any "unfair" way, we encourage the readers to study their highly interesting publications, contributed to the community both within and without the VISORSURF project scope.

CHAPTER 2

Introduction

Christos Liaskos, Ageliki Tsioliaridou, Sotiris Ioannidis

Foundation for Research and Technology Hellas, 71110, Heraklion, Crete, Greece

Andreas Pitsillides, Ian Fuat Akyildiz

Computer Science Department, University of Cyprus

What if the laws of electromagnetism could be customized, rather than being simply adhered to? In recent years, a novel research direction has emerged to explore this direction in the context of wireless communications, exhibiting paradigm-shifting potential, especially in the much anticipated extremely high frequency regime. The present book explores this new direction at an educational level. We hope that the material, though covering multiple disciplines, is fit for newcomers in the field.

The book revolves around artificial materials which can sense electromagnetic waves impinging upon them and alter these waves in ways dictated by software. These materials, such as the recently proposed HyperSurfaces [132] concept, can coat common objects such as walls, doors, and sizeable furniture, exerting logical control over the way wireless propagation occurs within an environment overall, e.g., within a building. Thus, the environmental behavior can be dynamically optimized to the needs of wireless devices, such as mobile phones and laptops, offering unprecedented levels of communication range, data transfer rates, energy efficiency, and security [103].

A core component of HyperSurfaces are the metasurfaces, i.e., planar, artificial structures, which have recently enabled the realization of novel objects with engineered and even *unnatural* electromagnetic functionalities [89]. Metasurfaces can be designed for any operating frequency and application domain. HyperSurfaces constitute a kind of networked metasurfaces, which can connect to the Internet of Things ecosystem and provide wireless environment optimization under any context [58]. The metasurface functionalities can then be described in software, such as parametric wave steering, absorbing, filtering, and polarizing [16, 99, 101].

The derived practical benefits are highly promising. As shown in Fig. 2.1, interconnected HyperSurfaces deployed within a space can mitigate previously unsurmountable, degenerative factors in wireless propagation—namely, the path loss, the multi-path fading, and the Doppler shift. This readily allows for lower-power transmissions, which favor the battery lifetime of the wireless devices. Moreover, the decreased scattering reduces cross-device interference, allowing an increased number of mobile users to co-exist in the same space, without degrading their performance. Additionally, the traveling wave reaches the receiver via well-defined paths rather than via multiple echoes, allowing for increased data transmission rates and high-quality coverage even at previously "hidden" areas. From another aspect, this separation of user devices can target increased privacy. Waves carrying sensitive data can be tuned to avoid all other devices apart from the intended recipient, hindering eavesdropping. These interesting environmental behaviors and more can be expressed in software in the form of combinable and reusable modules. Thus, communication system designers and operators are enabled to easily and jointly optimize the complete data delivery process, including the wireless environment, supplementing the customizable wireless device behavior, and, furthermore, reducing the complexity of the device design.

HyperSurfaces stand on the verge of several scientific disciplines, including Materials Physics, Electrical Engineering, Manufacturing of Electronics, Communications, and Computer Science. This book aspires to provide a basic inter-disciplinary scientific background for newcomers, to delve into the resulting new research direction. To this end, the chapter structure is as follows:

Chapter 3 provides the foundational concepts of electromagnetism that permeate and describe the operation of metasurfaces and metamaterials.

Chapter 4 models the metasurface foundations from the Computer Science aspect, introducing novel software classes that enable the transparent integration of HyperSurfaces into the Internet of Things ecosystem.

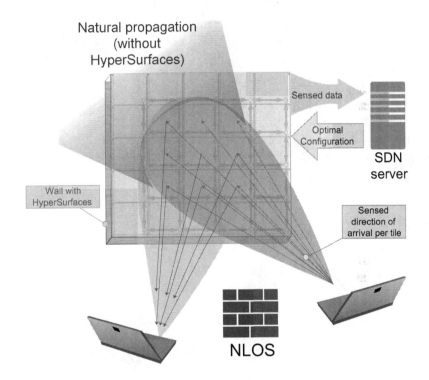

Figure 2.1 General schematic on how a deployment of HyperSurfaces can lead to software-defined wireless propagation within a space.

Chapter 5 details the network protocols that interconnect the HyperSurfaces and their components to the external world, providing access to their functionalities at the network level.

Chapter 6 details prospectful Internet of Things-compliant approaches for inter-networking metamaterials. Existing solutions in the IoT market are surveyed, from the aspects of hardware, software, and standards.

In Chapter 7, we make a short interim and provide a categorization of the preceding and ensuing chapters in a form of an early protocol stack, in order to facilitate exposition.

Chapter 8 studies the scaling laws of HyperSurfaces, i.e., how their performance, cost, and energy expenditure is scaled versus various factors.

Chapter 9 showcases the Internet of Metamaterials applied to the wireless communications case, with a focus on studying their inter-networking approaches.

Finally, Chapter 10 concludes the book by outlining promising future directions.

CHAPTER 3

Electromagnetic Specifications and Prototype Designs of Software-Defined Surfaces

Fu Liu, Xuchen Wang, Mohammad Sajjad Mirmoosa, Sergei Tretyakov

Aalto University, Department of Electronics and Nanoengineering, Espoo, Finland

Odysseas Tsilipakos, Anna C. Tasolamprou, Maria Kafesaki, Alexandros Pitilakis, Nikolaos V. Kantartzis

Foundation for Research and Technology Hellas, Heraklion, Crete, Greece

Do-Hoon Kwon

University of Massachusetts Amherst, Department of Electrical and Computer Engineering, Amherst, Massachusetts, USA

CONTENTS

3.1	Electromagnetic Modeling of Metasurfaces	10
	3.1.1 Unit Cell, Polarizability, and Interaction Constant	10
	3.1.2 Impedance Boundary Condition	13
	3.1.3 Sheet Impedance	15
3.2	Metasurfaces and Their Functionalities Compared with Other Sheet Materials and Phased Array Antennas	18
	3.2.1 Sheets of Usual Materials	19
	3.2.2 Antenna Arrays	20
	3.2.3 Metasurfaces	22
	3.2.4 Comparison of Possible Functionalities and Advantages of Metasurfaces	22

3.3   Tunable Metasurfaces: From Global Tuning to
       Software-Defined Metasurfaces 25
       3.3.1   Global Tuning 25
                3.3.1.1   Electric Tuning 26
                3.3.1.2   Magnetic Tuning 27
                3.3.1.3   Tuning by Light 27
                3.3.1.4   Thermal Tuning 27
       3.3.2   Local Tuning 28
                3.3.2.1   Switch Diode 29
                3.3.2.2   Continuous Tuning Varactors 29
                3.3.2.3   Coding Metasurfaces 30
       3.3.3   Software-Defined Metasurfaces 31
3.4   Design Workflow of the Switch-Fabric Prototype 32
       3.4.1   Design Principles per Functionality 32
                3.4.1.1   Tunable Perfect Absorption 33
                3.4.1.2   Tunable Anomalous Reflection 34
       3.4.2   Proof-of-Concept Design and Its Performance 36
                3.4.2.1   Tunable Perfect Absorption 36
                3.4.2.2   Tunable Anomalous Reflection 37
       3.4.3   Practical Considerations and Restrictions 40
                3.4.3.1   Electronic Package Considerations
                           from the EM Aspect 42
                3.4.3.2   Where Should We Position the
                           Package Vertically? 43
                3.4.3.3   Linking Electromagnetic and Package
                           Designs 44
                3.4.3.4   Basic Unit-Cell Parameters 49
                3.4.3.5   Advanced Design Considerations 50
       3.4.4   Overview of the Switch-Fabric Tunable Absorber
                Design 50
3.5   Electromagnetic Performance of the Switch-Fabric
       Design 51
       3.5.1   Tunable Perfect Absorption 51
       3.5.2   Anomalous Reflection 53
       3.5.3   Polarization Conversion 55
       3.5.4   Electromagnetic Characterization Procedures 57
                3.5.4.1   Experimental Setup 57
                3.5.4.2   Demonstration Procedure 59
                3.5.4.3   Spatial Modulation of Load
                           Configuration for Better
                           Performance 61

3.6   Design of the Graphene-Based Prototype 62
       3.6.1   Practical Considerations and Design
                Constraints 62
       3.6.2   Graphene Combined with Metallic Patches 67
       3.6.3   Frequency-Tunable Perfect Absorber 70
       3.6.4   Switchable Absorber 72
       3.6.5   All-Angle Perfect Absorber 73
3.7   Summary 75

A metasurface is an electrically thin composite material layer, designed and optimized to function as a tool to control and transform electromagnetic waves [55,63,78,167,192]. The layer thickness is small and can be considered as negligible with respect to the wavelength in the surrounding space. Recently, the emerging topic of tunable metasurfaces for wave control [38, 40, 45, 68, 69, 82, 106, 119, 177] has been actively advanced thanks to new possibilities for versatile and powerful control over the propagation of electromagnetic (EM) waves. Among the many available tuning mechanisms, the lumped-element enabled tunable metasurface provides the most powerful functionalities. However, the existing solutions are only using diodes or varactors, which tune only the effective surface impedance in a limited way, such as discrete values or a line in the complex impedance plane. A more general solution is to control both the real and imaginary parts of the surface impedance, i.e., tuning both the R and C parts of tuning components.

This book discusses the design, fabrication, implementation, and performances of a tunable metasurface that can perfectly absorb the incoming electromagnetic wave from different incidence angles, realize anomalous reflection, and control polarization of reflected waves. In this chapter, we present the electromagnetic design of the unit cell for the tunable metasurface. The content of this chapter is as follows. Sections 3.1 to 3.3 present basic information on metasurfaces and tunable metasurfaces. We discuss the effective-parameter modeling of metasurfaces and their electromagnetic properties in Section 3.1, compare metasurfaces with other sheet materials in Section 3.2, and review the tuning mechanisms in Section 3.3. Then, in Sections 3.4 and 3.5, we discuss the design flow of a switch-fabric prototype and its electromagnetic performance. In Section 3.6, we present designs of graphene-based prototypes. And, finally, we summarize the chapter in Section 3.7.

3.1 ELECTROMAGNETIC MODELING OF METASURFACES

In this section we present electromagnetic models of generic metasurfaces and basic information about their electromagnetic properties.

3.1.1 Unit Cell, Polarizability, and Interaction Constant

We consider probably the simplest metasurface: just one layer of small electrically polarizable unit cells (elements), as illustrated in Fig. 3.1. There is no ground plane, no substrate, and the array elements are in free space. The array is periodical and infinite, and the period is smaller than half wavelength. Moreover, we assume that the unit cells are electrically small and each unit cell can be considered as an electric dipole. We denote the dipole moment of one unit cell as $\mathbf{p}$. Let us also assume that the metasurface is planar and the unit cells are planar (metal) patches in the plane of the metasurfaces. In this case all dipole moments of all unit cells are oriented in the plane of the metasurface.

The dipole moment of a unit cell is proportional to the electric field at the point where this particular unit cell is located, measured in the absence of this unit-cell patch. The electric field which excites each unit cell is called **local field** $\mathbf{E}_{\text{loc}}$. Assuming linearity, for each unit cell of the metasurface we can write

$$\mathbf{p} = \alpha \mathbf{E}_{\text{loc}}, \tag{3.1}$$

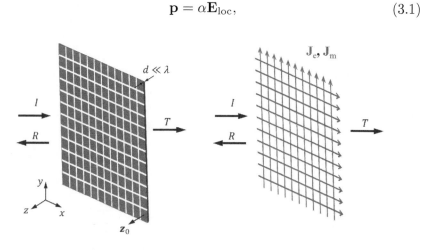

Figure 3.1 Upon surface averaging of currents in unit cells, the metasurface can be considered as a sheet of electric and magnetic surface currents. In the simple example presented in this picture, the unit cells are only electrically polarizable, and the induced magnetic current $\mathbf{J}_{\text{m}} = 0$.

where parameter α is called **polarizability**. The polarizability of one unit cell (one dipole particle) can be computed or measured if we take one single unit cell out of the array and study it isolated (in free space). The local field at the position of a particular unit cell is the sum of the incident field and the fields scattered by all other elements in the array. This model is illustrated in Fig. 3.2.

Let us consider excitation of metasurfaces by external fields. For example, we illuminate the array by a plane wave, whose field we call the **incident field** $\mathbf{E}_{\mathrm{inc}}$. If we would know how dipole moments of unit cells depend on the incident field, that is, if we would know the coefficient $\hat{\alpha}$ (called **collective polarizability**) in the relation

$$\mathbf{p} = \hat{\alpha}\mathbf{E}_{\mathrm{inc}}, \tag{3.2}$$

we would be able to find the reflection and transmission coefficients in terms of $\hat{\alpha}$ and then control them varying $\hat{\alpha}$. The concept of the individual and collective polarizabilities is illustrated in Fig. 3.3.

Let us simplify our problem even further and assume that the excitation is a normally incident plane wave. In this case all the unit cells feel the same exciting field, and all the dipole moments are the same. Now, assuming that we know $\hat{\alpha}$, we can find the induced electric surface current density $\mathbf{J}$. Indeed, the surface density of electric polarization is simply the dipole moment per unit area, and we find it by dividing the dipole moment of one unit cell $\mathbf{p}$ by the unit-cell area S. Next, we note that the time derivative of the polarization density vector is the surface current density, which gives

$$\mathbf{J} = \frac{j\omega\mathbf{p}}{S} = \frac{j\omega\hat{\alpha}}{S}\mathbf{E}_{\mathrm{inc}}. \tag{3.3}$$

We note that this relation has different physical meaning than the differential Ohm's law, because Ohm's law relates the current density and

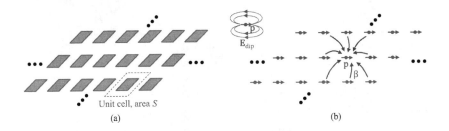

Figure 3.2 Periodically arranged unit cells (a) and their model as interacting electric dipoles (b).

the *total* (not the *incident*) electric field. A two-dimensional analogue of Ohm's law is derived in Section 3.1.2.

This electric current sheet radiates plane waves in both forward and back directions, and the amplitude of the electric field created by this electric current sheet is

$$\mathbf{E} = -\frac{\eta}{2}\mathbf{J}, \tag{3.4}$$

where η is the free-space impedance. This result is obtained by equating the circulation of magnetic field around a current source to the current, which gives $2H = J$, and the relation between electric and magnetic fields in a plane wave, $E = \eta H$. Then, the reflection coefficient R of the metasurface reads

$$R = \frac{E}{E_{\text{inc}}} = -\frac{j\omega\eta}{2S}\hat{\alpha}. \tag{3.5}$$

However, we do not know the collective polarizability $\hat{\alpha}$, because the unit cells are excited by the local field, which is not equal to the incident field! The local field is the sum of the incident field and the field created by the currents induced in all the other unit cells, which is called the **interaction field $\mathbf{E}_{\text{int}}$**, as illustrated in Fig. 3.3(a).

In general, we can write Eq. (3.1) as

$$\mathbf{p} = \alpha(\mathbf{E}_{\text{inc}} + \mathbf{E}_{\text{int}}). \tag{3.6}$$

If we know the incident field and the properties of one unit cell, we still need to find the interaction field, in order to find out what will be the metasurface response to this excitation. In the assumption of

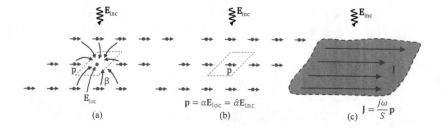

Figure 3.3 (a) The local field is the sum of the incident field and the interaction field. (b) The unit-cell dipole moment can be expressed in terms of the individual polarizability α or the collective polarizability $\hat{\alpha}$. (c) The plane-wave reflection and transmission properties are defined by the surface-averaged electric current density.

excitation by a normally incident plane wave, all the dipoles are the same, and the interaction field is proportional to the dipole moment (of each unit cell) $\mathbf{p}$:

$$\mathbf{E}_{\text{int}} = \beta\mathbf{p}. \tag{3.7}$$

Here, parameter β is called the **interaction constant**. It depends on the array period, the frequency, the properties of the surrounding space (the presence of a substrate, for instance), but it does not depend on the unit cell polarizability α.

If we know the interaction constant β, we can find the induced dipole moment in terms of the incident field. Combining Eqs. (3.6) and (3.7), we find

$$\mathbf{p} = \alpha(\mathbf{E}_{\text{inc}} + \beta\mathbf{p}), \qquad \mathbf{p} = \frac{\alpha}{1 - \alpha\beta}\mathbf{E}_{\text{inc}} = \hat{\alpha}\mathbf{E}_{\text{inc}}. \tag{3.8}$$

Now the problem is solved, since we know the reflection coefficient in terms of the unit-cell properties, and we can control it by tuning the polarizablity of unit cells.

We note that this approach works only for plane-wave excitations of infinite periodical (uniform) arrays. For normal incidence on an infinite array all the dipoles are the same, and we can write (3.7). For oblique incidence of plane waves all the dipoles have the same amplitude, and we know the phase shift from one unit cell to the next one. We can still use Eq. (3.7), where $\mathbf{p}$ is the dipole moment of one particular unit cell, which is at the origin of the coordinate system. In this case, calculations become more difficult, but still possible. However, as soon as the excitation is not a single plane wave or the array has a finite size (or it is not periodical), we need to solve the problem globally, because all the dipoles are in general different and we just have a linear system of equations containing all unknown dipole moments of each unit cell, which all interact with each other. In fact, if the array size is finite but large (compared to the period), an infinite-array approximation works well, because only 2-3 edge unit cells will "feel" the array edge.

3.1.2 Impedance Boundary Condition

On the other hand, metasurfaces can be modeled by some effective parameters, as we do for usual materials and for metamaterials using permittivity and permeability, which do not depend on excitation or the size and shape of the sample.

Let us first find the impedance (resistance) per square of our simple metasurface. For a thin sheet of some material which can carry electric current (polarization or conduction current, it does not matter), **sheet impedance** is defined as the ratio of the tangential electric field on the sheet plane and the surface current density:

$$E_{\text{total}} = Z_{\text{g}} J. \tag{3.9}$$

Here, we use scalars assuming isotropy: the current is flowing in the same direction as the electric field which drives it. The surface current density J is the same as above. The total electric field E_{total} is the sum of the incident field E_{inc} and the surface-averaged field E which, given by Eq. (3.4), is created by the currents induced in the particles. Both quantities are averaged over several unit cells, as illustrated in Fig. 3.4, so that the averaging area is still small compared with the wavelength. One can also say that E_{total} equals E_{loc} plus the field created by the dipole at the 0 position, but both these fields must be averaged over the unit-cell area.

One can envisage a square sample of the size 1 m times 1 m, which contains many unit cells, so that we can speak about averaged currents and fields. Then the voltage across this square equals E_{total}, as the electric field is measured in volt per meter, and the sample is one meter long. In the same way, the surface current in the sample equals J, as the sample is one meter wide and J is the surface current density with the unit of ampere per meter. That is why Z_{g}, measured in Ohms, is sometimes called impedance per square. Index "g" is used to indicate that in our case it is the impedance of a *grid* of unit cells. Note that in (3.9) the field is surface-averaged over the unit cell area: grid impedance is an effective parameter of the homogenized metasurface. Of course, also the surface current density is the averaged surface current.

If we know the sheet impedance, we can find the reflection coefficient in terms of Z_{g}, and we can control it by tuning Z_{g}. For example, for normally incident plane waves, the reflection coefficient reads [166,

(a) (b)

Figure 3.4 Impedance boundary condition (3.9) relates surface-averaged surface electric current density and the surface-averaged tangential electric field.

Eq. (4.101)]

$$R = -\frac{\eta/2}{\eta/2 + Z_g}. \tag{3.10}$$

For $Z_g \to 0$ (ideally conducting sheet) we have $R \to -1$, for $Z_g \to \infty$ (no induced current) we have $R \to 0$, etc.

Let us again consider our simple example of an array of electric dipoles excited by a normally incident plane wave. As it was already discussed, the electric field in the metasurface plane E_{total} is the sum of the incident field and the surface-averaged field created by the induced dipole moments. In the array plane, this secondary field varies a lot, but upon surface-averaging we have the field of the plane wave created by the whole array. Thus, we can write

$$E_{total} = E_{inc} - \frac{\eta}{2}J, \tag{3.11}$$

where we have used Eq. (3.4). Note that this field is the same on both sides of the array, because we have only electric polarization in this simple example. Thus, the transmission coefficient obeys the relation $T = 1 + R$.

In order to find the impedance per square Z_g in (3.9), we need to "get rid of" E_{inc} in the above equation. We can do it using Eqs. (3.3) and (3.8). Indeed, by combining these two equations we have

$$J = \frac{j\omega}{S}\frac{\alpha}{1-\alpha\beta}E_{inc}, \quad \text{or} \quad E_{inc} = \frac{S}{j\omega}\left(\frac{1}{\alpha} - \beta\right)J. \tag{3.12}$$

Substituting E_{inc} into Eq. (3.11), we find the impedance sheet condition in the desired form (3.9):

$$E_{total} = \left[\frac{S}{j\omega}\left(\frac{1}{\alpha} - \beta\right) - \frac{\eta}{2}\right]J = Z_g J. \tag{3.13}$$

This result gives the effective (homogenized) sheet impedance as a function of the polarizability α of a single (located in free space) unit cell. With the use of this formula, one can study how the effective properties of the array and the reflection and transmission coefficients depend on the individually tunable parameters of a single unit cell.

3.1.3 Sheet Impedance

Now we see how the effective sheet impedance ("resistance per square") depends on the parameters of individual unit cells α and the interaction constant β:

$$Z_{\mathrm{g}} = \frac{S}{j\omega}\left(\frac{1}{\alpha} - \beta\right) - \frac{\eta}{2}. \tag{3.14}$$

In general, the sheet impedance is a complex number, whose real part (resistance per square) determines dissipation loss, and whose imaginary part tells about surface reactance. For instance, for dense arrays of disconnected metal patches the imaginary part is negative, corresponding to the surface-averaged capacitance of the array.

Let us first consider our simplest example of an electrically dense periodical and infinite array excited by a normally incident plane wave. Let us recall the definition of the interaction constant β. This parameter defines the interaction field in terms of the dipole moments of unit-cell particles, as given in Eq. (3.7) or the induced surface current density:

$$E_{\mathrm{int}} = \beta p = \beta\frac{S}{j\omega}J, \tag{3.15}$$

where we have used Eq. (3.3). But what is the physical meaning of this interaction field? We take an infinite array of identical dipoles p, remove one dipole, and calculate the field created by all the remaining (still identical) dipoles in this, now empty, unit cell. This is the interaction field which we add to the incident field to find the field which will excite this one dipole when we will bring it back to its place. So, the surface-averaged value of this interaction field is just the plane wave generated by the whole array (except the particle at the observation point). It is equal to

$$E_{\mathrm{plane\ wave}} = -\frac{\eta}{2}J, \tag{3.16}$$

which is from Eq. (3.4), minus the field of one single dipole. This simple consideration defines the interaction constant due to the surface-averaged interaction field:

$$\beta_{\mathrm{aver}} = -j\frac{\eta}{2}\frac{\omega}{S} + j\frac{k^3}{6\pi\epsilon_0}, \tag{3.17}$$

where $k = \omega\sqrt{\epsilon_0\mu_0}$ is the free-space wavenumber. A detailed derivation and discussion of this important formula can be found in [166, Sec. 4.5.2]. The first term is the plane-wave field, whose negative imaginary quantity corresponds to power radiation from the array, and the second term is the single-dipole field, with a positive imaginary quantity because it is what we subtract from the plane-wave field of the complete infinite array. Note that the inverse polarizability of any electric dipole particle has exactly the same term to account for scattering loss (radiation damping) [166, Eq. (4.82)]. For a lossless unit cell made from a perfect conductor or a lossless dielectric, the polarizability is anyway a complex number because the particle loses power on radiation:

$$\frac{1}{\alpha} = \operatorname{Re}\left(\frac{1}{\alpha}\right) + j\frac{k^3}{6\pi\epsilon_0}. \tag{3.18}$$

Equation (3.17) tells about unit-cell interactions via the surface-averaged fields, which are plane waves carrying energy from the metasurface. But the array particles are not uniformly filling the unit cell area. Rather, there are small dipoles sitting in the middle of every cell. Thus, we need to account for non-uniformity of the interaction field inside the unit cell. These fast-varying fields are evanescent (they do not propagate away and do not carry energy), thus, they do not change the imaginary part of the interaction constant, given in Eq. (3.17). Instead, they define the real part of the interaction constant. Thus, in general, for regular infinite arrays of small dipoles we have a complex interaction constant

$$\beta = \operatorname{Re}(\beta) - j\frac{\eta}{2}\frac{\omega}{S} + j\frac{k^3}{6\pi\epsilon_0}. \tag{3.19}$$

Now we can substitute the interaction constant in Eq. (3.19) into the formula for the sheet impedance in Eq. (3.14) and find, for lossless unit cells,

$$Z_{\mathrm{g}} = -j\frac{S}{\omega}\left[\operatorname{Re}\left(\frac{1}{\alpha}\right) - \operatorname{Re}(\beta)\right]. \tag{3.20}$$

We see that all real parts (meaning scattering loss) in the expression for the impedance cancel out, and the result is a purely imaginary impedance, meaning that the array is a reactive sheet characterized by surface-averaged capacitance. If particles in the unit cells are lossy (dissipative), we have an additional resistive term:

$$Z_{\mathrm{g}} = -j\frac{S}{\omega}\left[\operatorname{Re}\left(\frac{1}{\alpha}\right) - \operatorname{Re}(\beta)\right] + \frac{S}{\omega}\operatorname{Im}\left(\frac{1}{\alpha}\Big|_{\mathrm{diss}}\right). \tag{3.21}$$

The subscript "diss" highlights that here the imaginary part of the polarizability accounts only for dissipation, and the scattering loss (radiation damping) term in Eq. (3.18) is excluded, since it has been already accounted for.

3.2 METASURFACES AND THEIR FUNCTIONALITIES COMPARED WITH OTHER SHEET MATERIALS AND PHASED ARRAY ANTENNAS

Tunable metasurfaces are thin (in comparison with the wavelength of operation) layers whose electromagnetic properties can be engineered and dynamically reconfigured. Because there are other known possibilities to realize such reconfigurable thin layers, it is important to understand the differences and similarities of various approaches. The most critical issue is to identify and use advantages of the metasurface technology compared to conventional and already established technical solutions.

Three basic scenarios of realizing engineered and tunable electromagnetic "skins" are illustrated in Fig. 3.5. Case (a) corresponds to a very thin sheet made of an ordinary material whose properties can be dynamically adjusted. For instance, a rubber sheet or a semiconductor layer (controlled, e.g., by carrier injection or by incident light) or a ferrite layer, controlled by a bias magnetic field, etc. Here, the "array elements" are atoms or molecules, so the array period is very small as compared to the wavelength even in the visible. Case (c) corresponds to a phased array antenna. Here, the array elements are antennas (any kind of antenna can be used, including patch antennas). Finally, case (b) is the metasurface, where the distance between the array elements is not negligibly small, but significantly smaller than the wavelength (typically, about $\lambda/5\ldots\lambda/10$).

All these three types of tunable sheets can realize an angle-tunable absorber for plane waves or tilt reflected waves. To realize a tunable absorber, a tunable sheet is positioned above a ground plane. The sheet resistance and reactance (or the antenna input impedance for the

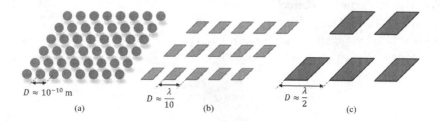

$D \approx 10^{-10}\,\text{m}$ $D \approx \dfrac{\lambda}{10}$ $D \approx \dfrac{\lambda}{2}$

(a)   (b)   (c)

Figure 3.5 Conceptual illustration of a sheet of a usual material (a), a metasurface (b), and an antenna array or a frequency selective surface (c). D is the array period, and λ is the operational wavelength.

phased array realization), required for full absorption of plane waves of a given frequency, polarization, and incidence angle, are known from the literature and can be realized in various ways. In the following we will discuss the three scenarios and what possible advantages and more complex functionalities one can possibly get using the metasurface approach.

3.2.1 Sheets of Usual Materials

Let us first consider the limiting case of sheets made of usual materials. For instance, a very thin metal sheet or a sheet of paper. Each molecule is a tiny electric dipole, and we can use the theory of dipole arrays to understand the properties of such material sheets (well, assuming that it is so thin that we have only one layer of molecules, but it is not a critical limitation for understanding).

In this scenario, the array period is extremely small compared to the wavelength. We can consider a small (compared to the wavelength) area of the sheet and for every molecule within this small piece the environment would look the same as if the sheet would be infinite and uniformly excited. Indeed, the incident field is the same for millions of molecules around any reference molecule. In this situation, we can characterize the sheet by conventional macroscopic parameters, such as resistance per square or, more generally, the sheet impedance. In addition, the result in Eq. (3.19), which we obtained by assuming that the array is infinite and uniformly illuminated, will be valid for arbitrary external field configurations (for any incidence angle or for spherical-wave illumination). Furthermore, the real part of the interaction constant in Eqs. (3.19) and (3.21), which is defined by how fields vary between two neighbouring dipoles in the array, can be calculated using the quasistatic approximation, because the distance between the molecules is so small. The static equations of electromagnetics physically mean the assumption that electromagnetic interactions take place instantaneously, without any time delay. This model works well if the unit cell is so small that the propagation time over its volume can be neglected. In this situation, the spatial field distributions can be found from electrostatics, and then we can assume that this spatial field profile harmonically oscillates in time, with the same phase at every point inside the unit cell. Assuming that the molecules are point dipoles and retaining only the static terms in the expression for the dipole fields, a simple calculation (see details in [166, Fig. 4.14]) gives

$$\mathrm{Re}(\beta) \approx \frac{0.36}{a^3 \epsilon_0}, \tag{3.22}$$

where a is the array period, i.e., the distance between molecules. $1/a^3$ dependence corresponds to the leading term in the near field of a dipole, that decays as $1/r^3$. Note that this quantity does not depend on the incidence angle and even on the frequency! Thus, in this limit of very dense arrays of very small unit cells, the sheet impedance is indeed a simple effective parameter, which depends only on the "material" (the polarizability of one molecule α and on the distance between them a). If we know Z_{g}, we can use it to find the response to any external fields, and the sample can be of a finite size or curved.

Tuning of the sheet properties is made by tuning lots of array elements (molecules) together, this way changing the value of the sheet impedance. For instance, using ferrites or piezoelectrics, or mechanically changing the sheet such as like rubber stretching.

3.2.2 Antenna Arrays

The case of antenna arrays corresponds to the other extreme, where the unit-cell size is comparable with the wavelength (usually equals half wavelength) and the number of elements is not huge (usually from tens to hundreds, except large military radars). The effective-medium description in terms of the sheet resistance is based on the assumption that the interaction field created by neighboring antennas, averaged over one unit cell, is a plane wave of an infinite and uniformly excited array. Obviously, this assumption for obtaining Eq. (3.17) is not valid at all in this case. Moreover, the distribution of reactive near fields inside one unit cell is rather complicated and it depends on the phase and amplitude difference between currents in the neighbouring unit cells. Reactive (evanescent) fields store electromagnetic energy and do not carry it away from the array elements. The static point-dipole estimation in Eq. (3.22) which assumes that the particles are identically excited point dipoles is not applicable at all. The near-field distribution over one unit cell now depends on the array excitation (the incidence angle) and on the frequency. Basically, the introduction of surface-averaged fields and currents as well as the very notion of the sheet impedance (3.9) loose sense for antenna arrays. In this case, every array element is controlled individually and considered as an antenna. No homogenization is possible. One needs to control each antenna element, taking into account coupling between the array elements which depends on the array excitation. Tuning the array properties is made by careful and individual tuning of each array element, usually varying loads of antennas which form the array.

The typical array period of $\lambda/2$ is chosen as the largest period still allowing scanning the direction of transmission or reception/absorption over the whole half space. When the elements are fed in phase, the radiated field interferes constructively in the normal direction. If the neighbours are out-of-phase, the constructive interference is in the direction along the surface. Varying the phase gradient between these two extremes, one can scan from -90 to $+90$ degrees in both planes. The same functionality is obviously available for smaller periods, but then the cost increases with increasing the number of array elements.

Lots of results about phased arrays are known, tunable arrays (called phased array antennas) are widely used, and there are also adaptive (self-tuning) antenna arrays, which make use of clever signal processing algorithms... The most common adaptive functionality is to maximize the received power for signals coming from a certain direction while cancelling interference from other directions (see an illustration in Fig. 3.6). In these systems, the amplitude and phase in each unit cell are adaptively tuned so that the array pattern takes the optimal shape. More advanced systems are capable of adaptive reception from several users at the same time. Early results on adaptive arrays can be seen, e.g., in [14, 180].

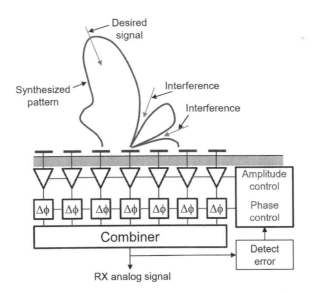

Figure 3.6 Conceptual illustration of an adaptive antenna array.

3.2.3 Metasurfaces

Usually, metasurfaces use several (many but not so many, say five to ten) elements per one wavelength. And the overall number of unit cells is large but not huge. In this scenario, introducing the homogenized model of the sheet resistance and impedance, i.e., Z_g in Eq. (3.9), still makes sense, but its value is now a function of the incident field configuration. This is because in this case we cannot assume that each element is surrounded by a large number of identical elements which are all identically excited. When the incidence angle changes, there will be some phase difference between the currents in different cells, and it will be rather significant. One can perhaps say that we have a mesoscopic structure, intermediate between the usual materials and antenna arrays.

For plane-wave illumination of infinite and periodic metasurfaces it is possible to calculate the interaction field between unit cells and, thus, the sheet impedance numerically or even analytically. It is possible because the unit cells differ only by the phase of the induced currents, and this phase difference is known if we know the angle of incidence. In terms of simulations, it means that it is enough to simulate just one unit cell in periodical boundary conditions. However, if the array is finite or elements are not exactly the same (which is the case of practically interesting metasurfaces for reflection control!) or distances between them are not all the same, or the illumination is not a plane wave, it is not possible. We have a global problem for the whole array, where all unit cells interact with each other. For realizing anomalous reflection from an electrically large metasurface we can model it as a set of periodically arranged supercells, but even in this case each supercell contains many small unit cells, that are all different and must be optimized together.

3.2.4 Comparison of Possible Functionalities and Advantages of Meta-surfaces

Suppose that the required functionality of our device is such that the "input field" is a single plane wave and the "output field" is another plane wave traveling in the same direction or reflected specularly (possibly having zero amplitude in the case of a perfect absorber [134]). To realize such response, the surface should be uniform. If it is an array of unit cells, then the period should be smaller than $\lambda/2$ and all the cells should be the same, at least when the response is averaged over any wavelength-size area. Here we assume that the array size is electrically large and neglects the edge effects.

To realize such simple functions, one can use either of the three classes of tunable surfaces: material sheets (e.g., mechanically stretchable) or phase arrays, or metasurfaces. The choice can be made based on availability, simplicity, and cost of the control element (just one control signal is needed). It is perhaps advantageous to use conventional phased-array antennas, just because here the number of unit cells is minimized. On the other hand, since the bias voltage is the same for all units, it perhaps does not matter much how many of them we have. From the practical point of view, it is perhaps better to have several units per wavelength because it can be easier to set the average properties to the correct values by having several step-wise controlled units rather than just one.

On the other hand, if the functionality which we want to realize is more complex, so that the surface should *change* the wave propagation direction, then the surface should be non-uniform on the wavelength scale. Here, homogeneous material sheets become practically problematic as it is not easy to control material properties in a wide range and maintain good spatial resolution, especially at high frequencies. Thus, antenna arrays or metasurfaces are used. If the desired function is "not too extreme", such as anomalous reflection into moderate angles or focusing with a large focal distance, both conventional antenna arrays and metasurfaces can do the job well. However, antenna arrays exhibit more significant "discretization errors" of the required phase distribution over the surface, since the unit cell is larger. The effects of this error are pretty small and usually can be neglected. However, if the functionality is more demanding, for example, anomalous reflection into very steep angles or focusing with a small, comparable to λ, focal distance, metasurfaces can offer functionalities which antenna arrays cannot realize.

To better illustrate these differences, let us consider the anomalous reflection from surfaces made of unit cells with different sizes d_x. For simplicity, we consider two-dimensional wave propagation on the $x - y$ plane and the surface is along the x-direction. Further, we assume normal incidence. With different numbers of unit cells, a supercell with size D_x can be formed to enable the anomalous reflection to the angle

$$\theta_r = \arcsin \frac{m\lambda}{D_x}, \tag{3.23}$$

where λ is the wavelength of the incoming wave, and m is the diffraction order. Ideally, when the supercell is made of extremely small units, i.e.,

$d_x \to 0$, it can be modeled with a continuous surface impedance Z_s, which is defined as the ratio of the tangential components of the total electric and magnetic fields. If the surface impedance function follows

$$Z_s(x) = j\eta \cot[\Phi_r(x)/2], \tag{3.24}$$

with the phase function $\Phi_r(x)$ spans $2m\pi$ phase linearly, the surface can effectively reflect the incident wave into the desired anomalous direction with efficiency [47–49]

$$\eta = \frac{4\cos\theta_r}{(1 + \cos\theta_r)^2}. \tag{3.25}$$

However, the unit-cell size cannot be arbitrarily small, and in practice the surface impedance will be discretized. As an example, in Fig. 3.7 we compare the performance of surfaces made of unit cells with sizes $\lambda/2$, $\lambda/5$, and $\lambda/10$, which represent the phased array or metasurface. As we can see, when the unit cell size is $d_x = \lambda/10$, the calculated efficiencies are close to the analytical results given by the black line. However, when the unit-cell size is $d_x = \lambda/2$, the efficiency drops noticeably for large anomalous reflection angles.

We stress that if one would desire to have full control over the reflected field, the use of metasurfaces is the only possibility, as conventional phase arrays are not versatile enough.

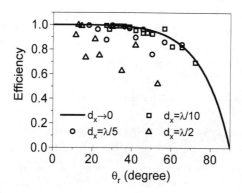

Figure 3.7 Efficiency of anomalous reflection for metasurfaces with different sizes of unit cells.

3.3 TUNABLE METASURFACES: FROM GLOBAL TUNING TO SOFTWARE-DEFINED METASURFACES

There are many tuning mechanisms which can be used to realize tunable metasurfaces. In this section, we review these mechanisms, from global tuning to local tuning, and then to the software-defined metasurfaces.

3.3.1 Global Tuning

A metasurface can be made globally tunable by introducing in the unit cell (meta-atom) composition stimuli-responsive materials, which are capable of undergoing relatively large and rapid changes in their physical properties in response to external ambient stimuli [148]. In effect, when the ambient conditions change (e.g., temperature, pressure, humidity, applied electric/magnetic field or pump light beam), then the material properties will be tuned accordingly and therefore modify the metasurface response. As the ambient stimuli apply to the

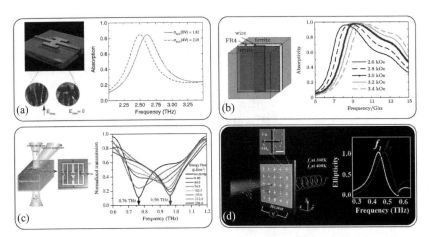

Figure 3.8   Tunable metasurfaces with stimuli-sensitive materials. (a) Electric-sensitive liquid crystal-based tunable metasurface absorber. Reprinted with permission from [147]. Copyright 2013 by the American Physical Society. (b) Magnetic-sensitive ferrite-based tunable metasurface absorber. Reprinted with permission from [191]. (c) Light-sensitive semiconductor-based tunable metasurface. Reprinted with permission from [145]. Copyright 2011 by the American Physical Society. (d) Thermal-sensitive VO_2-based tunable quarter-wave plate. Reproduced from [176] under Creative Commons CC BY license.

entire metasurface area, we refer to this approach as "global tuning". In what follows, we will focus on the tunability by electric and magnetic fields, light, and temperature stimuli, with some characteristic examples presented in Fig. 3.8.

3.3.1.1 Electric Tuning

Nematic liquid crystals (LCs) give a famous example of electrically sensitive materials, due to the massive development of the optical display technology. Nematic LC molecules respond to the bias electric field by rearranging their orientation and thus achieving voltage dependent birefringence. Due to their liquid nature, they can be infiltrated into various metasurface structures providing large refractive index modulation for operation in the microwave, terahertz and optical regime [50, 121, 195, 204]. To date, nematic LCs have been used in various metasurface applications such as tunable absorbers [147], Fig. 3.8(a), and cloaks with possibilities of real-time control of invisibility [125]. Due to their large anisotropy, LCs have also been employed in hyperbolic metasurfaces [31].

Another famous electrically sensitive material is graphene, whose Fermi level can be modified by external electrostatic field. For example, in [71], researchers have demonstrated tunable plasmon resonances in graphene microribbon arrays on a silicon/silica substrate. However, the weak interaction between graphene and light, which is due to poor obtainable mobility (especially for processed graphene), hinders the accomplishment of practical functionalities. In fact, from an equivalent circuit perspective, the weak interaction comes from the huge impedance mismatch between graphene and the surrounding materials. While low-quality graphene has very large surface impedance in the terahertz range, the characteristic impedance of the surrounding materials is comparably low, limiting the tunability of the total metasurface impedance. To circumvent this issue of low tunability, metallic inclusions are introduced in conjunction with graphene to significantly reduce the effective surface impedance. In this way, the graphene-metal hybrid metasurfaces have shown efficient tunability of the reflection amplitude, phase, and resonance frequency [46, 70, 81, 87, 178]. For example, a recent work in [81] shows that a graphene-metal strip structure can demonstrate strong tunable absorption (with absorbance modulation efficiency of 96%) when the Fermi level is tuned from 0.26 eV to 0.57 eV.

3.3.1.2 Magnetic Tuning

Magnetically tunable structures are attractive thanks to their instantaneous response to external contactless magnetic stimuli. Via magnetic field-guided self-assembly of colloidal particles, the induced dipole-dipole interactions between these particles can be accurately controlled (attractive or repulsive). As a result, by properly adjusting the controlling magnetic field profile the assembly process can be tuned, providing means to control the resulting response. Applications include adaptive microplate arrays, i.e., magnetically responsive nanostructures for tuned optics [111], hybrid devices [123] and generation of resonant-assisted tunneling phenomena [27]. Magnetically responsive structures have been exploited also in tunable-resonance metamaterials: ferrite-wire metamaterials for tunable negative index [61], ferrite-rod structures with different saturation magnetization for wide-band tunable microwave filters [24], and tunable broadband absorbers consisting of ferrite slabs and copper wires [191], as illustrated in Fig. 3.8(b).

3.3.1.3 Tuning by Light

Photoconductive semiconductor materials, such as Si and GaAs, have been employed for implementing tunable metasurfaces by tuning their conductivity through carrier photoexcitation by an infrared pump beam. Several implementations with split-ring-resonator-based metasurfaces operating in the THz have been proposed. In paper [33, 122], semiconductor (GaAs) is included as the substrate material. Alternatively, semiconductor (Si) can be incorporated as a section in the resonant structure [35, 145]. In this case, carrier photoexcitation effectively modifies the resonator geometry. This can lead to a redshift of the fundamental resonance [35] or trigger a transition to a different, blue-shifted resonance [145], as shown in Fig. 3.8(c). Semiconductor inclusions have been utilized also in chiral metasurfaces for switching their handedness [198] or tuning the dichroism and optical activity [75].

3.3.1.4 Thermal Tuning

Heat is another ambient stimulus for tunable metasurfaces. Meta-molecules or substrates whose macroscopic parameters are sensitive to temperature variations can lead to the change of the electromagnetic response of metasurfaces. In this context, phase change materials (PCMs) with temperature-dependent permittivity are promising candidates [52, 53, 79, 143, 146, 175, 176]. For instance, vanadium dioxide, as one of the well-known PCMs, behaves as an insulator at room temperature and transforms to a metal state at higher temperatures due to the enhancement of the free carrier concentration. It has been employed in tunable metasurfaces to shift the resonance frequency, modulate the transmission amplitude [52, 53] or switch the polarization of the transmitted wave at two different frequencies [176], as shown in Fig. 3.8(d).

3.3.2 Local Tuning

While global tuning applies to the entire metasurface as a whole, local tuning is a powerful approach that provides a possibility to tune the properties of each unit cell individually. In this way, we can achieve additional tunable functions, such as steering, focusing, imaging, and holography [37, 67, 90, 108, 169]. For local tuning we can use the same stimuli-sensitive materials; however, in this case we need to apply the stimuli locally in each unit cell. For example, a heater/cooler is needed in each unit to effectively control the temperature, a coil (capacitor) in each unit to change the local magnetic (electric) field, and a light emitting diode in each unit to apply different illuminations [144]. These, however, are usually impractical solutions due to additive large objects compared to the unit cell size even in the GHz-range metasurfaces. A practical option is to use voltage-driven elements, for example diodes and varactors, which have relatively small sizes and can be effectively addressed. Figure 3.9 lists some examples with different levels of complexity.

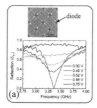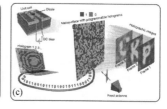

Figure 3.9 Varactor-loaded tunable metasurfaces with increasing complexity and ability. (a) A diode switches the functionality from total absorption to total reflection. Reprinted from [202], with the permission of AIP publishing. (b) Column-controlled varactors enable a tunable metasurface to perform multiple functions: splitting, steering, and polarization conversion. Reproduced from [67] with permission. Copyright Wiley-VCH Verlag GmbH & Co. KGaA. (c) Patterned coding-metasurface enables dynamic hologram creation. Reproduced from [90] under Creative Commons Attribution 4.0 International License (http://creativecommons.org/licenses/by/4.0/).

3.3.2.1 Switch Diode

A switch diode, which has two states "on" and "off", enables dual-function metasurfaces in the GHz band, by applying appropriate control voltages on the diodes. For example, Fig. 3.9(a) shows one design in which total absorption and total reflection can be switched by changing the state of the diodes [202]. In this case, the diodes modify the input impedance of the metasurface to match (total absorption) or mismatch (total reflection) with the free-space impedance. In another design [159], both the polarization and scattering properties can be modified. While the incident linearly polarized wave is reflected to the same polarization when the diodes are on, the same incident wave will be transmitted with perfect polarization rotation when the diodes are switched off.

3.3.2.2 Continuous Tuning Varactors

Varactor diodes (variable capacitors) can be adjusted in a continuous way [199]. In the simplest scenario, all varactors are controlled by one and the same voltage, which effectively gives frequency tunability to the functionality that the metasurface is designed for [30, 51, 115, 117]. The most widely investigated functionality is tunable perfect absorption, where a change in the reverse biasing voltages of the varactors shifts the spectrum of the perfect absorption resonance [80, 112, 114, 203]. Typical values for the (reverse) voltage range are 0–20 V corresponding to an equivalent capacitance range of a few pico-Farad, e.g., 0.5–3.5 pF, for commercially available diodes that are compact enough to be integrated in GHz-band metasurfaces. The corresponding frequency tunability accessible by this capacitance range is in the order of a couple of GHz, e.g., 4–6 GHz. Similarly, microelectromechanical systems (MEMS), in which the capacitances are tuned by the piezo-electric effect, provide another option for continuously tunable high impedance metasurfaces [38].

Moving one step towards more elaborate functionality, the locally-applied continuous tuning voltage is allowed to be different for each unit cell of the metasurface. For example, a voltage profile can be applied on the varactors in one direction while keeping the voltage unchanged in the other direction, as shown in Fig. 3.9(b). In this way, a specific one-dimensional phase profile can be actively "imprinted" on the metasurface. This enables a range of tunable applications, such as tunable reflection (steering) [39, 67, 149, 185], beam splitting [67], and even writing a letter by dynamically changing the focal point with a tunable "Huygens' metasurface" [37]. Evidently, the price to pay for this enhanced functionality is the increased complexity in the electrical network that controls the voltages biasing the varactor diodes of the metasurface.

On the other hand, instead of modulating the unit cell properties only in one direction, the metasurface can be dynamically programed into full two-dimensional patterns. In this case, all individual varactors can be controlled independently and therefore one can obtain arbitrary distributions, which have enabled researchers to demonstrate dynamical holograms [90], as shown in Fig. 3.9(c). Evidently, the electrical network of such a programmable metasurface increases in complexity, since $N \times N$ diodes need to be biased. As a compromise between functionality and complexity, one can use two diodes in each unit cell and connect one set of diodes column-wise and the other set row-wise. By doing so, only $2 \times N$ control voltages are required, while the programmable metasurface can still give promising functionalities, for example, single-sensor imaging by utilizing many available configuration patterns [91].

3.3.2.3 Coding Metasurfaces

An interesting approach for obtaining quasi-continuous control via discrete states consists in clustering unit cells with a binary response in groups of two, three, or more (N), forming composite building blocks (which should remain subwavelength in size) that can offer 2^N surface impedance states. A similar approach is based on embedding more diodes in each unit cell maintaining individual control over each of them. These approaches have been termed as *coding* and *digital* metasurfaces [45]. The metasurface states are being built by *encoding* a discrete set of unit-cell options that can be seen as *bits* or *words*, and the implementation in the GHz is typically via digital control through reconfigurable logic, e.g., a Field-Programmable Gate Array (FPGA) [45]. Coding metasurfaces provide a powerful and intuitive design perspective, while drawing a clear parallelism with information theory and opening new ways to model, compose, and design advanced metasurfaces.

This concept has been exemplified in several works with a variety of functionalities, including beam focusing [65], wavefront manipulation [201], angular momentum conversion, and polarization control [197].

3.3.3 Software-Defined Metasurfaces

While research of metasurfaces has exhibited impressive capabilities in controlling EM radiation in every physical aspect (amplitude, phase, frequency, direction, polarization) and in a wide range of frequency regimes (microwave and THz to infrared and optical), their incorporation in real-world systems is still not straightforward. Recently, a new research direction has sought to push metasurfaces into real-world, massive applicability. This direction is termed as software-defined metasurfaces (SDMs) or HyperSurfaces (HSFs), and is detailed in subsequent chapters. Here we briefly outline the concept for the physics inclined audience.

SDMs and HSFs seek to act as the linchpin between academic research and real-world applicability of metamaterials and metasurfaces, by harnessing the opportunities offered by i) a well-defined software interface and ii) an integrated nanocontroller network that allows communication between the unit cells and with the outside world [103].

The realization of a software-defined metasurface requires a hardware system that applies software primitives and effectively reconfigures the metasurface. To this end, some researchers propose to integrate a network of tiny controllers within the metasurface structure and to wirelessly interface it with an external entity [104]. Each controller is capable of interpreting global instructions and of acting locally by, for instance, tuning its varactors to achieve the desired impedance configuration. The main challenges here are, firstly, to develop an ultra-low

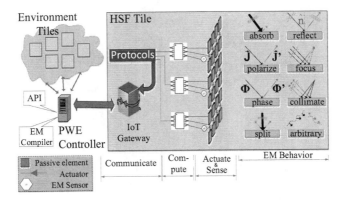

Figure 3.10   Inter-operating layers comprising a HyperSurface. 2019 IEEE. Reprinted, with permission, from [103].

cost network of controllers and, secondly, to co-integrate it with the metasurface within a single structure.

The SDM architecture is accordingly organized in three functionality layers, shown in Fig. 3.10: the intra-networking layer, the gateway layer, and the software control layer. Several specific implementations of SDMs can take place, depending on the components and architectures chosen to implement each layer.

3.4 DESIGN WORKFLOW OF THE SWITCH-FABRIC PROTOTYPE

The target is to realize a tunable metasurface that is capable of tunable perfect absorption for different incidence angles and tunable anomalous reflection. This section discusses the design flow of the switch-fabric prototype and the specifications of the finalized design.

3.4.1 Design Principles per Functionality

First of all, we describe the theory behind the tunable perfect absorption and the tunable anomalous reflection [108], as the two examples of tunable functions which can be given by a tunable metasurface structure. In order to understand the theory, at this point let us consider a simple structure which we employ later to show the numerical results corresponding to those tunable functions. Figure 3.11 illus-

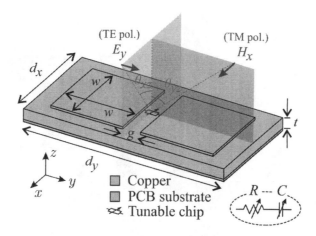

Figure 3.11   Schematic of the unit cell for the proposed metasurface. The dimensions are listed in text. A tunable ICs package is incorporated to provide a variable complex impedance and locally modify the surface impedance of the metasurface in a continuous way.

trates the schematic view of one unit cell of the proposed structure. As seen, the structure is a periodical array of metal patches positioned on a thin grounded dielectric substrate. We stress that the proposed unit cell can be realized for operating at different frequencies. However, since our goal is to work at 5 GHz, we suggest the following values for the unit cell properties. The dielectric substrate possesses the thickness of $t = 1.016$ mm and the size of each rectangular unit cell is $d_x = d_y/2 = 9.12$ mm. The square patches have the width $w = 8.12$ mm (metallization thickness 17.5 μm) and are separated from each other by a 1-mm gap where a tunable integrated circuit (IC) is connected. The thin dielectric substrate is a high-frequency laminate Rogers RT/Duroid 5880 which is characterized by the relative permittivity $\varepsilon_r = 2.2(1 - j\tan\delta)$ with the dissipation factor $\tan\delta = 0.0009$.

Such metasurface structure in Fig. 3.11 can provide us with multiple tunable functions at 5 GHz, i.e., tunable perfect absorption for different incidence angles and tunable anomalous reflection toward different directions. For that, as mentioned above, there is a tunable integrated circuit (IC) which is introduced into the unit cell and connects the two metal patches along the y direction, as shown in Fig. 3.11. The IC is modeled as continuously tunable, lumped complex impedance loads, with a tunable capacitance C (negative reactance) and a tunable resistance R. We assume that the tuning range is approximately $0 - 5\ \Omega$ for R and $1 - 5$ pF for C.

3.4.1.1 Tunable Perfect Absorption

The copper background on the back side of the metasurface (ground plane) prevents transmission of the incident wave. Therefore, the performance of the metasurface is characterized by the reflection coefficient. Consequently, we only need to minimize the reflection coefficient in order to obtain the perfect absorption [134]. To do that, the input impedance seen at the metasurface plane should be matched to the free-space impedance. However, as the characteristic impedance of plane waves in vacuum is varying for different incidence angles θ_i and for different polarizations, i.e.,

$$Z_{0,\text{TE}} = \frac{\eta}{\cos\theta_i}, \quad Z_{0,\text{TM}} = \eta\cos\theta_i, \tag{3.26}$$

one needs to tune the input impedance to achieve perfect absorption when the incident angle changes.

The input impedance is the surface impedance of the sheet (coming from the gaps between patches and the IC contribution) in parallel with the impedance seen from the substrate. From the transmission-line theory point of view, since the substrate is electrically thin and it is grounded, the impedance seen from the substrate is similar to a reactance of an inductance. As a consequence, it can be understood that the capacitance C provided by the IC determines the resonance frequency which should happen at the operation frequency of 5 GHz, and the resistance R (provided by the IC) controls the absorption level. Since the change of the incident polarization and incident angle changes the required input impedance for perfect absorption, by proper tuning of the reactance and resistance, we readily achieve the tunability characteristic for perfect absorption. Although the required input impedance remains purely resistive for any incidence, it is necessary to tune both surface resistance and reactance of the patch array. This tuning can be done by properly adjusting the properties of the metasurface globally and continuously, i.e., all unit cells must have the same configuration, but the allowed settings must not be limited to a discrete set of values.

3.4.1.2 Tunable Anomalous Reflection

Anomalous reflection is achieved if the impinging wave is deflected away from the specular direction [20, 32, 47, 86, 133, 135, 181, 193]. Therefore, to obtain such an intriguing feature, we should construct a supercell of N unit cells (the total extent is $D = Nd$) with varying capacitances $C_1, C_2, \ldots, C_N$. In Fig. 3.12(a), a supercell along the x axis is depicted that can perform anomalous reflection in the $x - z$ plane. Changing the supercell size enables anomalous reflection to different directions. For example, when the supercell size D is $2\lambda_0 < D < 3\lambda_0$ and under normal incidence, four diffraction orders ($m = \pm 1, \pm 2$) besides the specular ($m = 0$) are propagating. The reflection angle for a given diffraction order, m, is determined by momentum conservation [108]

$$k_0 \sin \theta_r = k_0 \sin \theta_i + m(2\pi/D), \qquad (3.27)$$

in which k_0 is the free-space wavenumber and θ_r denotes the reflection angle. From the above equation, one concludes that

$$\theta_r = \arcsin(m\lambda_0/D), \qquad (3.28)$$

for normal incidence ($\theta_i = 0$). Note that with different sizes of the supercell, promoting different diffraction orders allows us to achieve a quasicontinuous coverage for the reflection angle. This feature is illustrated in Fig. 3.12(c) where we plot the supported anomalous reflection angles for the first and second diffraction orders as the number of cells in the supercell varies. It is worth mentioning that the phase should have a linear profile along the supercell [108]. In other words,

$$\phi(x) = \phi_0 - m(2\pi/D)x. \tag{3.29}$$

By this way, we can determine a specific diffraction order over the remaining leakage channels. We also note that the resistance R of the ICs is purposely set to zero for all unit cells to minimize absorption.

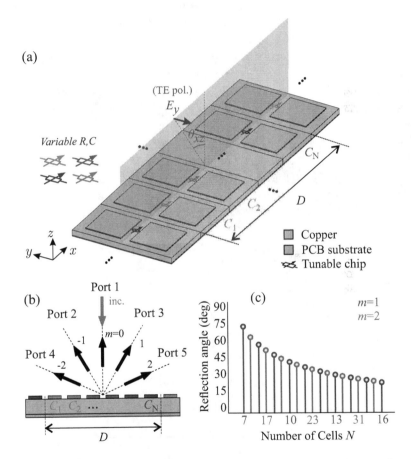

Figure 3.12   (a) Schematic view of the supercell for anomalous reflection in the $x - z$ plane. Different shades of the lumped elements represent different capacitance settings. (b) Port naming convention and correspondence with reflected diffraction orders for normal incidence. (c) Anomalous reflection angles for different numbers of cells exploiting the first and second diffraction orders (assuming normal incidence).

3.4.2 Proof-of-Concept Design and Its Performance

In the following, we give the corresponding results of tunable perfect absorption and tunable anomalous reflection, for the proof-of-concept design. The specifications of the unit cell have been described in the previous subsection.

3.4.2.1 Tunable Perfect Absorption

We investigate the tunable perfect absorption feature for different incidence angles and two main polarizations: Transverse-electric polarization (electric field is in the y direction and the incidence plane is the $x - z$ plane), and transverse-magnetic polarization (magnetic field is parallel to the x axis and the incidence plane is the $y - z$ plane), as illustrated by Fig. 3.11. To find the required capacitance and resistance values provided by the ICs (series topology as shown in the inset of Fig. 3.11) corresponding to perfect absorption at 5 GHz, we follow a numerical approach and simulate the structure. For normal incidence in which two polarizations are the same, we find from simulations that we need $R = 3.44 \ \Omega$ and $C = 1.69$ pF to fully absorb the incoming wave. For oblique incidence, varying the incidence angle from 10° to 70° for both polarizations, we also obtain the required R and C val-

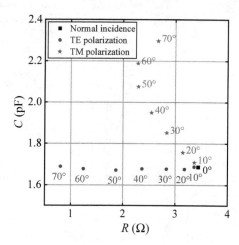

Figure 3.13 The required R and C values of the electronic elements (series topology, see inset in Fig. 3.11) for achieving perfect absorption at different incidence angles for the two orthogonal polarizations (see Fig. 3.11). The operation frequency is 5 GHz.

ues described by Fig. 3.13. As it is seen, for the TE polarization, the required capacitance does not change dramatically, and it is approximately $C \approx 1.68$ pF. However, the required resistance varies from 0.8 Ω (corresponding to 70°) to 3.4 Ω (corresponding to 10°). On the other hand, for the TM polarization, the situation is different. Both R and C experience remarkable changes as the incident angle varies. The different TE/TM behavior originates from the different impedance matching conditions for oblique incidence angles. Regarding the capacitance, we can think about the specific case where the structure does not have any ICs. In that case, Fig. 3.13 confirms that the reactive part of the input impedance for the TM polarization is very sensitive to the angle of incidence as compared to the TE polarization. That is why for the TE polarization the required value of the capacitance for perfect absorption does not noticeably vary with respect to the angle of incidence. About the resistance, it is clear because when the incidence angle changes, the free-space wave impedance changes differently for the two polarizations. This means that different resistance modulations are required for the two polarizations.

3.4.2.2 Tunable Anomalous Reflection

In order to specify the required capacitances for the initial linear-phase prescription, we need a "look-up table" relating the reflection phase with the capacitance of the load. It is specified by illuminating the uniform metasurface with a normally incident plane wave at the operation frequency of 5 GHz, and the results are shown in Fig. 3.14. To be realistic, we limit the achievable series capacitances in the range $[1, 5]$ pF. Even under this restriction, we have access to a large reflection phase span of 300°, while the reflection amplitude remains almost unity (inset).

Transverse-electric polarization: As the first example, we consider a supercell consisting of $N = 8$ unit cells stacked along the x axis (notice that $D = Nd_x = 72.96$ mm). For the normal incidence we get $m = \pm 1$ diffraction orders in the $\pm 55.3°$ directions. The supercell can be, thus, effectively described by a three-port network, as shown in Fig. 3.15(c). Using the look-up table in Fig. 3.14 we specify the capacitances of the lumped loads for achieving a linear phase profile and promoting the $m = 1$ diffraction order; they are depicted in Fig. 3.15(a). The corresponding discrete reflection phases are shown in the inset. Notice that the first and last points deviate slightly from the prescription of a linear reflection phase. This fact occurs because we limit the available capacitance range to $[1, 5]$ pF. Using the capacitance values given in Fig. 3.15(a) we characterize the supercell response by successively exciting the three ports and extracting the reflection coefficients in each port (S parameters). The entire S-parameter matrix is (absolute values of the components in decibels)

$$|S|_{\text{dB}}^{\text{initial}} = \begin{bmatrix} -26.65 & -12.55 & -0.68 \\ -12.55 & -2.01 & -10.15 \\ -0.68 & -10.15 & -28.36 \end{bmatrix}. \tag{3.30}$$

Notice that the matrix is symmetric as dictated by reciprocity. For normal incidence (port 1), the incident power is indeed mostly reflected to the $m = 1$ diffraction order (port 3), as can be seen by the high amplitude of the S_{31} element in Eq. (3.30). However, there is some unwanted radiation in port 2 ($m = -1$) as well ($|S_{21}| = -12.55$ dB). Parasitic reflections are anticipated since i) the periodic approximation under which the look-up table of Fig. 3.14 is obtained is not exact when we switch from the unit cell to the supercell and ii) there is a nonzero variation of the local reflection amplitude along with the local reflection phase. These are the known limitations of the "phase-gradient" approach. Here, we use optimization to achieve perfect anomalous reflection. Specifically, we seek capacitances that maximize $|S_{31}|$ (the optimization goal is set to $|S_{31}| > -0.5$ dB), while at the same time minimizing both $|S_{21}|$ and $|S_{11}|$ (the optimization goal is set to $|S_{21}|, |S_{11}| < -20$ dB). The optimized capacitance values are

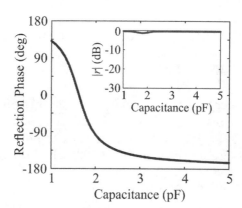

Figure 3.14 Look-up table: Reflection phase for a uniform metasurface as a function of the capacitance of the tunable IC. The operating frequency is 5 GHz and the incident wave impinges at normal incidence. The corresponding reflection amplitude is shown in the inset.

plotted in Fig. 3.15(b) and the obtained S-parameter matrix for the optimized metasurface reads:

$$|S|_{\mathrm{dB}}^{\mathrm{optimal}} = \begin{bmatrix} -20.17 & -20.50 & -0.47 \\ -20.50 & -1.23 & -16.11 \\ -0.47 & -16.11 & -18.39 \end{bmatrix}. \tag{3.31}$$

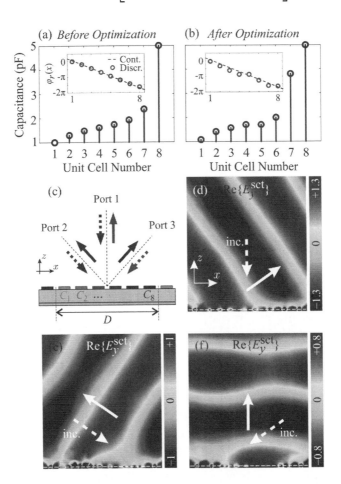

Figure 3.15 Anomalous reflection with an $N = 8$ supercell promoting the first diffraction order. (a),(b) Capacitance values for the meta-atoms before optimization (linear phase prescription) and after optimization. The corresponding phase profiles are shown in insets. (c) Port naming convention and the three reflection scenarios: $1 \to 3$, $2 \to 2$, $3 \to 1$. (d),(e),(f) Scattered electric field for the three reflection scenarios. Planar wavefronts tilted to the desired direction are observed.

As we can observe, with the optimized supercell, we increase the reflection towards port 3 and suppress radiation towards port 2, thus achieving nearly perfect anomalous reflection. This result is lucidly illustrated in Fig. 3.15(d) where we plot the scattered electric field: nicely planar wavefronts tilted toward port 3 are observed, with negligible parasitic reflections.

Although optimization is performed for normal incidence, we improve the metasurface performance for incidence from ports 2 and 3 as well. Excitation from port 3 leads to reflection to port 1 with a high amplitude of $|S_{13}| = |S_{31}| = -0.47$ dB (reciprocity) and excitation from port 2 leads to retroreflection with a high amplitude of $|S_{22}| = -1.23$ dB. The above property can be clearly seen in Fig. 3.15(e),(f), where the scattered electric field is depicted for the two cases showing minimal parasitic reflections.

Transverse-magnetic polarization: The unit cells can be readily rearranged by varying the tunable impedances of the unit-cell loads. This property enables us to get perfect anomalous reflection also for the TM polarization ($\mathbf{H} = H_x \hat{\mathbf{x}}$) in the $y - z$ plane. In this case, we construct supercells along the y axis [Fig. 3.16(a)] to realize the required phase gradient along the y direction. We consider the case of $N = 5$ (note that $D = Nd_y = 91.2$ mm) giving rise to ± 1 diffraction orders toward $\pm 41.1°$.

Again, we start with the prescription of a linear phase profile and subsequently use optimization to fine-tune the load capacitances and the corresponding phase distribution [Fig. 3.16(b)]. The resulting S-parameter matrices before and after optimization are

$$|S|_{dB}^{initial} = \begin{bmatrix} -15.72 & -9.18 & -1.02 \\ -9.18 & -2.01 & -7.70 \\ -1.02 & -7.70 & -34.66 \end{bmatrix}, \qquad (3.32)$$

$$|S|_{dB}^{optimal} = \begin{bmatrix} -27.66 & -25.48 & -0.32 \\ -25.48 & -0.65 & -20.33 \\ -0.32 & -20.33 & -25.70 \end{bmatrix}. \qquad (3.33)$$

Excellent performance is obtained through optimization. The scattered magnetic field when the metasurface is illuminated from port 1 is depicted in Fig. 3.16(c), showing perfect anomalous reflection toward $41.1°$ (port 3) in the $y - z$ plane.

3.4.3 Practical Considerations and Restrictions

In the previous sections, we showed that a simple proof-of-concept design can demonstrate tunable perfect absorption and tunable anomalous reflection functionalities, with the tunability provided by variable lumped elements. In this section, we will discuss some issues of realizing a practical tunable unit cell, with embedded tunable lumped elements. Since two kinds of tunability (tunable resistance and tunable capacitance) are required in our design to achieve 'multi-functionality', a single diode/varactor cannot do the work.

Therefore, we will assume the use of a packaged electronic circuit (which provides the tunable lumped elements) to realize the required tunable parameters. Here, we will consider only the EM specifications that such a package should support, without any details on the internal package structure, actual implementation, or simulated/measured response. We will employ the term *electronic package* or *package* to denote this difference in the remainder of this chapter. More practical details, designs, and results are presented in [131], from the theoretical

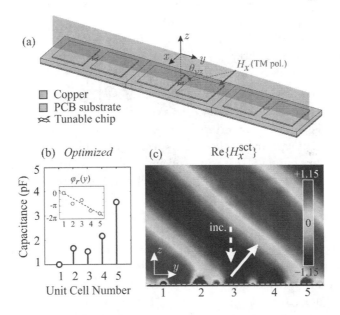

Figure 3.16 Anomalous reflection in the $y - z$ plane (TM polarization) with a $N = 5$ supercell along the y axis promoting the first diffraction order. (a) Schematic of unit-cell stacking along the y axis with $N = 3$. (b) Capacitance values for the meta-atoms after optimization; the corresponding phase profile is shown in the inset along with the linear phase profile prescription (dashed line). (c) Scattered magnetic field when excited from port 1: perfect anomalous reflection toward $41.1°$ (port 3).

EM-design perspective and basic practical considerations, and in [83], from the package-design perspective, also including advanced practical considerations and final simulated performance.

In the following subsections, we will discuss the basic configurations that the electronic package must provide, the various connections between the package ports and the patches of the unit cell. By briefly outlining the advantages and disadvantages of different layouts, we finally determine the best scenario, i.e., package below the ground plane and connected to four patches by means of through vias. In this way, we can decouple the EM design and package design, and use S-parameter block-modeling to connect the two designs and extract the final performance.

3.4.3.1 Electronic Package Considerations from the EM Aspect

The package is used to provide the tunable resistance and tunable capacitance values between pairs of patches, i.e., sets of $Z_{\text{in}} = R + 1/(j\omega C)$, where RC are assumed to be in series connection. Thus, the first step is to decide how we will connect the package to the patches.

Package-patch connectivity The proposed design builds upon the proof-of-concept design, essentially extending it to both lateral dimensions. This is achieved by adopting square unit cells, each one comprising 2×2 sets of square patches on its top layer. The four patches are all connected to a single electronic package which is situated in the center of the unit cell. The electronic package here is used for controlling the electromagnetic property of the unit cell and thus for the whole metasurface for a given incidence direction, polarization, and frequency. In this way, each package can host multiple lumped RC loads (e.g., two, four, or six) and fewer packages are required to serve the metasurface. As elaborated in [83, 131], while this has a generally negative effect on the metasurface performance from the EM point of view–due to increased package port number and, therefore, unavoidable cross-port power leakage–it facilitates manufacturability, since fewer packages are required and the assembly work and associated risks are minimized.

Tunable elements inside the package From an abstracted EM point of view, the tunable element connectivity inside the package can follow various alternatives. In our implementation [83, 131], each package contains four lumped RC loads, each one independently controlling the impedance presented by one pair of patches. The optimal internal connectivity of the four lumped loads inside the package is in an X-shape, with each load having one terminal near the inner corner of each patch and the other terminal at a common ground. This connectivity allows, in principle, full control of reflection (and absorption) over both polarizations and for any incoming plane wave direction (oblique and/or skewed).

In the above discussions, the tunable elements are all series RC loads, i.e., we work with impedance (resistance and reactance), but we can also connect them in parallel and/or work with admittance (conductance and susceptance). The conversion formulas between series and parallel RC loads, with respect to Q (quality factor) are the following:

$$Q = \frac{1}{\omega R_s C_s} = \omega R_p C_p, \tag{3.34}$$

$$\frac{R_p}{R_s} = 1 + Q^2, \tag{3.35}$$

$$\frac{C_p}{C_s} = \frac{Q^2}{1 + Q^2}. \tag{3.36}$$

Electronic package specifications from the EM aspect For the purpose of EM modeling of the package, and implementing the four RC loads in X-shape it delivers, we are only interested in five terminals: the four situated near the inner corners of the four patches of each cell, and one terminal in the middle of the cell. The positions of the four corner terminals and the dimensions of the respective copper pads are given in [83, 131].

3.4.3.2 Where Should We Position the Package Vertically?

Having defined the basic aspects of the lumped RC loads and their connectivity inside the electronic package from an EM perspective, we next consider where to place the package inside the unit cell and how to connect it to the patches. The three positions considered are depicted in Fig. 3.17; the proof-of-concept design uses the position in panel (a), while panel (b) depicts the loads embedded inside the top side dielectric substrate. For practical reasons related to manufacturability, serviceability, and EM performance, as outlined in [83, 131], these two options are dropped and the optimal placement of the electronic package is below the metasurface ground plane, as in Fig. 3.17(c). In this case, there are four vertical through vias (TVs) reaching up to the inner corners of each patch. A blind via (BV) can be used to short the middle terminal of the X-shaped four RC load group to the metasurface ground plane, but this is not required from an EM perspective.

3.4.3.3 Linking Electromagnetic and Package Designs

In the previous sub-subsection, we have chosen the configuration with the package below the ground as our preferential design, [83, 131]. The most important advantage of this configuration is the separation of electromagnetic (EM) and electronic package designs. In our case, the package contains four independent complex RC loads (in series connection) that connect the inner corners of the patches to a common reference terminal in the center of the cell, for symmetry. In this manner, the package effectively controls the behavior of the unit cell by tuning the values of these RC loads. For instance, for perfect absorption of an impinging plane wave of a given direction, polarization, and frequency, all unit cells of the metasurface (each one controlled by one package) must be tuned to the same values of resistance (R) and capacitance (C). For these optimal RC values, the metasurface presents a total surface impedance equal to the wave impedance of free space, i.e., 377 Ω, and thus the incident wave is totally absorbed without any reflection.

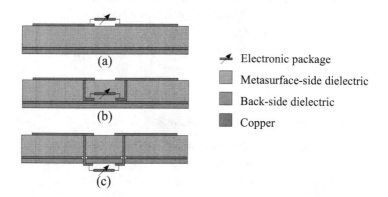

Figure 3.17 Cross sectional side-view of the unit cell configuration with the electronic package (consisting of lumped RC loads) (a) at the patches level, (b) embedded inside the top-side dielectric, (c) below the metasurface ground plane.

Ports for connecting the EM and package design. In the tasks dealing with the EM modeling of the unit cell, the aim is the design of its geometric and structural parameters and, then, the calculation of the lumped element values (i.e., the R and C values) that lead to the desired performance, namely: perfect absorption of an impinging plane wave of a given direction, polarization, and frequency. This procedure is performed without looking into the package structure as long as it produces the desired total impedance values at its ports (pair of terminals), e.g., $Z_{in} = R + 1/(j\omega C)$, if the RC loads are assumed here, without loss of generality, in series. Conversely, in the tasks dealing with the modeling of the package, the aim is to come up with a design that presents the desired total impedance values at its ports (pair of terminals), regardless of the unit cell EM performance. In order to minimize the interactions between these two design tracks, namely the EM modeling of the unit cell and the package modeling and architecture, it is imperative to effectively separate them. To that end, the four lumped elements implementing the package behavior are replaced by four discrete ports, each having one terminal near the inner corner of a patch and the other terminal on a common reference ground. Each package will henceforth be represented by a 4-port network fully described by the scattering parameters (S-parameters) it presents at its ports. Depending on the internal architecture of the package, it can be configured as four fully decoupled 1-port networks, two 2-port networks, or one 4-port network. In this work, we focus on the simplest (four 1-ports) configuration only, as the four RC loads are assumed entirely decoupled.

S-parameters for connecting the EM and package designs. The S-parameters of an N-port network are given in the form of an N-by-N matrix with complex (amplitude and phase) frequency-dependent values, referenced to a pre-specified characteristic (port) impedance (Z_0). A generalization of the reflection coefficient, the S_{ij} element of the matrix, is the ratio of the scattered wave voltage at the i-th port of the network over the incident wave voltage at its j-th port, assuming all the other ports are matched (terminated by loads with $Z_{in} = Z_0$). The S-parameters can easily be converted in any of the other families of parameters used to describe electrical, RF, and microwave networks, e.g., Z-, Y-, H-, T-, or ABCD-parameters. In simple 1-port networks, the complex impedance presented at its ports can be directly calculated as

$$Z_{in} = \frac{1 + S_{11}}{1 - S_{11}} Z_0 \Rightarrow S_{11} = \frac{Z_{in} - Z_0}{Z_{in} + Z_0}. \tag{3.37}$$

The ABCD parameters apply exclusively to 2-port networks and are 2-by-2 matrices. They can be used to easily extract expressions for networks composed of multiple cascaded loads (hence their other name: chain, cascade, or transmission parameters) by multiplying the respective 2-by-2 matrices of each load. Moreover, various lumped element circuits (e.g., series, parallel, T-, or Π-networks) have simple ABCD expressions. The following expressions are used to calculate the S-parameters of a 2-port network given its ABCD parameters:

$$
\begin{aligned}
S_{11} &= \frac{A + B/Z_0 - C/Z_0 - D}{A + B/Z_0 + C/Z_0 + D}, \\
S_{12} &= \frac{2(AD - BC)}{A + B/Z_0 + C/Z_0 + D}, \\
S_{21} &= \frac{2}{A + B/Z_0 + C/Z_0 + D}, \\
S_{22} &= \frac{-A + B/Z_0 - C/Z_0 + D}{A + B/Z_0 + C/Z_0 + D}.
\end{aligned}
\tag{3.38}
$$

In order to feed the extracted S-parameters of the package into the EM simulation of the unit cell, the Touchstone standardized representation format was chosen. Touchstone (or TS) files are plain ASCII files that contain the S-parameter data for the given N-port network in a column format: the first column is the frequency and then follow the columns for the amplitude and phase of each of the S-parameters $(1 + N^2$ columns in total). The extension of the filename contains the information of the number of ports of the network, e.g., S1P or S2P for 1- and 2-port networks, respectively.

Finally, note that the four lumped loads case (where Z_{in} is given) is in theory identical to the four 1-ports configuration (where S_{11} is given), due to the equality of Eq. (3.37), as long as the two terminals of the lumped loads/ports are identical: the one terminal is attached to a copper pad leading to a patch (with a TV) and the other terminal is attached to the common ground pad in the middle of the cell. The common ground pad in the middle of the cell can be grounded to the metasurface backplane by means of a blind via (BV), but is not imperative from the EM standpoint and can even deteriorate oblique TM-incidence performance. Figure 3.18 shows an example of how tuning the resistance or the capacitance of all the lumped loads (assumed identical) affects the spectral response of the unit cell reflection coefficient under normal incidence.

Block simulation using S-parameters. As outlined in [131], in order to efficiently model and simulate this multi-parametric EM problem, we rely on S-parameters, enabled by the metallic backplane, which effectively decouples the metasurface- and the package-side (tunable lumped RC load) designs. Thus, using the backplane as a common reference ground, we can model the package with four lumped ports, Fig. 3.19(a), and the incident and scattered plane waves with Floquet ports, Fig. 3.19(b). The lumped ports are 50 Ohm-referenced and fed by Touchstone data (complex spectra) depending on the package architecture, i.e., ranging from a single generalized four-port network (most complicated case) to four identical decoupled RC loads (simplest case). The Floquet ports are defined for a given plane wave direction, i.e., a (θ, ϕ) pair, both polarizations (TE and TM), and are placed at the front and back side of the unit cell, at planes perpendicular to the z-axis; periodic Bloch-Floquet conditions are applied to the remaining four boundaries enclosing the unit cell, emulating infinite periodicity. Please note that in the case of uniformly configured lumped loads/ports

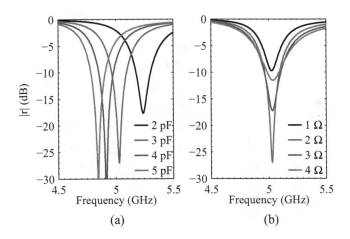

(a) (b)

Figure 3.18 Indicative spectral response of the unit cell reflection coefficient under normal incidence of a plane wave. The package was modelled by four identical 1-port networks or, equivalently, by four identical lumped RC loads. (a) For fixed $R_s = 2\ \Omega$, increasing the series capacitance shifts the near-zero reflection dip. (b) For fixed $C_s = 3$ pF, tuning the series resistance to a certain value, in this case close to $2\ \Omega$, can lead to zero reflection which corresponds to perfect absorption by the unit cell, assuming that the transmission to the back side of the cell is practically zero due to the ground plane.

we use a total of four Floquet ports: as the unit cell is subwavelength only the zero-order (specular) diffraction order is propagating and we in general need to consider both linear orthogonal polarizations. In the case of the inhomogeneous unit cell setting, where a 'supercell' with periodicity exceeding the free-space wavelength is formed, such as for the anomalous reflection scenario, the number of Floquet ports increases to accommodate the other propagating diffraction orders.

For the perfect absorber functionality, it is known that all unit cells of the metasurface must be identically configured; also, for our metal-backed isotropic structure, we can safely assume that transmission and cross-polarized coefficients will be practically zero. For these reasons, we model a single unit cell and we seek the package configuration that will lead to a minimization of the reflection coefficient for a given polarization, i.e., a minimization of the S_{55} or S_{66} scattering parameter, using the port-number convention of Fig. 3.19. As previously stated, the four lumped ports are decoupled, so that they can be directly replaced by lumped loads. This allows us to rapidly find the configuration, in terms of RC pair, that leads to perfect absorption at a given frequency and polarization: For a given incident plane wave direction, we numerically evaluate the 8-port S-parameters of the unit cell with full-3D EM simulation; then, we attach a prescribed RC load to all four lumped ports (passed as 1-port Touchstone data) and,

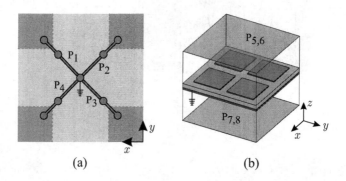

(a)   (b)

Figure 3.19 Modeling of the unit cell using combination of lumped and Floquet S-parameter ports: (a) Back side of the unit cell with its four lumped ports, $P_{1,2,3,4}$. (b) Perspective view of a subwavelength unit cell depicting the four Floquet ports, $P_{5,6}$ corresponding to TE and TM polarized incidence on the front side, and $P_{7,8}$ similarly on the back side. Floquet ports are defined on planes perpendicular to the z-axis, and for a given incident plane wave direction (θ, ϕ).

finally, we analytically extract the scattered spectra at the four Floquet ports, corresponding to the two polarizations (TE and TM) and to the two sides of the unit cell (top/back side, i.e., reflected/transmitted). These simulations are conducted in CST Microwave Studio (for full-3D) and its Design Studio module (for S-parameter block modeling). Iterating over the RC loads using an optimizer, we can rapidly identify the optimal configuration for perfect absorption at a given frequency and polarization. However, note that changing a geometric/EM parameter of the structure and/or the incident plane wave direction, e.g., for oblique incidence absorption, requires the numerical recalculation of the 8-port network S-parameters, i.e., a new full-3D EM simulation, which can be computationally intensive (time consuming) if the structure is electrically large and/or includes fine geometrical features.

3.4.3.4 Basic Unit-Cell Parameters

The unit cell design chosen is square (width w_{uc}) and contains four square patches (width w_p) symmetrically arranged in the 2×2 array in its top surface. As will be described in the following section, the width of the unit cell (approximately $\lambda/5$) and the patches greatly affect the performance and must be carefully tuned to the optimal values.

The dielectric chosen for the front and back side substrates, in accordance with fabrication partners, is Panasonic Megtron7N ($\varepsilon_r = 3.35$, $\tan\delta = 0.0020$). Laminate and prepreg layers were symmetrically positioned above and below the backplane in order to reduce PCB post-fabrication warping from asymmetric thermal expansion. From the availability of thicknesses of the specific dielectric materials, we concluded on a top-side total thickness (from patches to ground plane) of 2.14 mm and a back-side total thickness (from ground plane to package layer) of 0.26 mm.

The metallization layers are made of annealed copper with a thickness of 0.010 mm or 0.032 mm depending on the layer. The copper thickness does not visibly affect the response of the unit cell. According to the numbering of the metallization layers we adopted, [83], L1 is the patches layer, L2 is the ground plane (with holes for four TVs), and L4 is the layer where the lumped loads reside, where we only assume the traces from the lumped load outer pads (terminals) to the TV pads. Full patterning of the back-side metallization layers, which is required for other aspects of the metasurface operation, can be found in [83].

3.4.3.5 Advanced Design Considerations

Having decided upon the crucial structural parameters of the front side of the metasurface, we move on to the brief outlining of the equally important back side, where the package lies. More details and discussion on the design choices made can be found in [83, 131]. For the sake of completeness, we will only outline the aspects that need special attention:

- Position of the through vias (TVs) connecting the patches layer to the lumped loads layer: The optimal position is for the TVs to 'hit' the patches exactly on their inner corners. However, fabrication considerations may not allow that, so a new locally optimized position has to be identified; this typically comes with a compromise on performance, e.g., imperfect absorption for one polarization (geometric anisotropy).

- Modeling of the TVs: They are simulated as solid cylinders whereas in practice they are hollow cylinders; this does not have a crucial impact due to small skin depth at the operating frequency. Also, increasing the diameter of the TVs leads to a small increase in the absorption resonance frequency, without degrading its quality factor, so that perfect absorption can be restored by tuning the reactance (increasing the series capacitance).

- Position of the blind via (BV) shorting the pad/terminal in the center of the cell to the metasurface ground plane: Fabrication reasons might require moving the BV from the center of the cell; this has no effect on TE polarization, but can deteriorate oblique TM performance due to the E-field parallel to the via.

- Effect of solder balls connecting the copper pads to the actual electronic package: When solder balls are accounted for, a small shift in the optimal RC configuration is expected.

- Limitations on the RC range and bandwidth that the actual electronic package can provide effectively limit the oblique angle coverage for perfect absorption in both polarizations.

3.4.4 Overview of the Switch-Fabric Tunable Absorber Design

Concluding the tunable absorber design, we will recap on the technological limitations and constraints coming from package design and fabrication that led us to the final design, with details and dimensions presented in [83, 131]. Starting from elementary dimensioning considerations of the final HyperSurface design, we approximately quantify the size of the unit-cell and its type: Each cell has four square metallic patches in a 2×2 arrangement lying on an electrically thin metal-backed dielectric substrate; cell size is $\lambda/5$ or less; metal patches occupy more than half of the cell area. An even thinner back-side dielectric substrate, of equal type as the front-size dielectric, separates the backplane from the layer where the tunable lumped elements reside. Through vias (TVs) drilled through the substrates and the backplane are used to connect the lumped loads or the equivalent RF ports, to the metallic patches. Optimized performance (ample tuning range in terms of oblique incidence and operating frequency coverage) is achieved when the TVs strike the ground at the inner corners of the metallic patches, as close as possible to the outer terminals of the lumped loads. In [131], we outlined the methodology to automate the optimization of the TVs' positioning for any unit-cell and patch width, whereas more advanced considerations apply to other aspects of the final PCB fabrication and optimized performance; the latter is predominantly limited by the RC range and allowed bandwidth.

3.5 ELECTROMAGNETIC PERFORMANCE OF THE SWITCH-FABRIC DESIGN

In the previous section we have finalized the unit cell design of the switch-fabric prototype. In this section, we present the electromagnetic performance of the design and describe the measurement procedures to be followed for the experimental evaluation.

3.5.1 Tunable Perfect Absorption

First, we present the required settings of the complex load impedances for achieving perfect absorption at the working frequency of 5 GHz for both polarizations and variable incidence angles (oblique incidence). We use the carefully designed unit-cell design presented in the previous Section. The results have been obtained through an optimization procedure, by setting a minimization goal for the specular reflection of $|r| < -35$ dB. Since no higher diffraction orders are propagating (subwavelength periodicity) and transmission through the structure is almost prohibited due to the presence of the metallic backplane. This goal translates directly into perfect absorption.

Table 3.1 The required R and C to achieve perfect absorption for the TE polarization and for different incidence angles.

Angle (deg)	R(Ohm)	C(pF)
0	2.10	3.20
15	1.86	3.17
30	1.48	3.17
45	0.76	3.21
60	0.33	3.33
75	0.044	3.39

Table 3.2 The required R and C to achieve perfect absorption for the TM polarization and for different incidence angles.

Angle (deg)	R(Ohm)	C(pF)
0	2.10	3.20
15	2.11	3.215
30	2.53	3.31
45	3.15	3.42
60	4.19	4.01
75	3.62	7.17

More specifically, the required load impedances, modeled as a series connection of a resistor and a capacitor, are compiled in Table 3.1 for the TE polarization and Table 3.2 for the TM polarization, respectively. The considered incidence angles range up to 75 degrees in steps of 15 degrees. The configurations for studying TE and TM polarization are depicted in Fig. 3.20. Note that for TE polarization the incidence plane is the $x - z$ plane, as shown in Fig. 3.20(a), while for TM polarization the incidence plane is the $y - z$ plane, as shown in Fig. 3.20 (b).

The characteristic spectra of the reflection amplitude (in dB) demonstrating perfect absorption at the target operating frequency of 5 GHz are presented in Figure 3.21. More specifically, Figure 3.21(a) concerns the TE polarization and with an incidence angle of 30 degrees, while Figure 3.21(b) concerns the TM polarization for the same oblique incidence angle. In both cases, the optimization goal of $|r| < -35$ dB has been satisfied. The required load impedances, i.e., the R and C values in the package, as found by the optimization procedure, can be found in Table 3.1 for the TE polarization and Table 3.2 for the TM polarization, respectively.

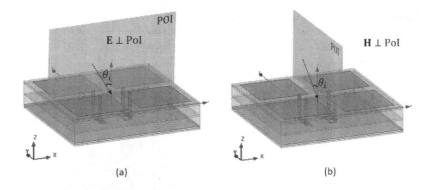

Figure 3.20 (a) Configuration for TE polarization oblique incidence; the plane of incidence (PoI) is the $x - z$ plane. (b) Configuration for TM polarization oblique incidence; the PoI is the $y - z$ plane.

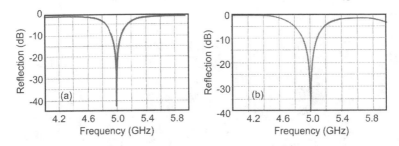

Figure 3.21 Characteristic spectra of the reflection amplitude (in dB) demonstrating perfect absorption at the operating frequency of 5 GHz. (a) TE polarization and 30 degree incidence. (b) TM polarization and 30 degree incidence.

3.5.2 Anomalous Reflection

The independent control over the RC values in each unit cell enables us to perform arbitrary wavefront manipulation, exemplified here via anomalous reflection. Within the realizable RC range by the package, although the reflection amplitude is not quite high, the coverage of the reflection phase is quite large, i.e., from -170 degrees to 170 degrees, as shown in Fig. 3.22. As a result, anomalous reflection can be achieved if we lower the requirement on the reflection amplitude. Anomalous reflection is obtained by grouping several unit cells into a supercell and allowing the reflection toward the desired direction while forbidding the scattering to other directions by selectively promoting a single propagating diffraction order over the others. Utilizing different supercell sizes and relying on the first or higher diffraction orders, the anomalous reflection direction can be tuned with a quasi-continuous angle coverage. To demonstrate the anomalous reflection effect, we consider a supercell made of $N = 8$ unit cells, which will support anomalous reflection from normal incidence toward the 56.4 degrees, according to Eq. 3.28 [108, 169], where D is the extent of the supercell and equals Nd_x for steering inside the $x - z$ plane for example. In this supercell configuration, there are three ports corresponding to the three propagating diffraction orders $m = 0, +1, -1$ (since $\lambda_0 < D < 2\lambda_0$). Note that transmission and polarization conversion is negligible and, thus, power collected by the corresponding ports is negligible. Subsequently, we perform an optimization while constraining the RC values within the realizable range. Note that in order to speed up the optimization process one can use as an initial setting for the RC values those dictated by a linear phase profile along the supercell, i.e., $\phi(x) = \phi_0 - (2\pi/D)x$ [108].

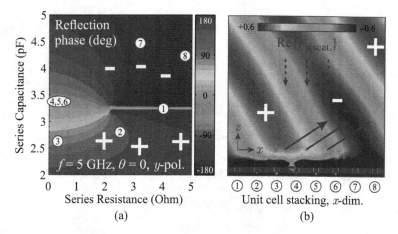

Figure 3.22 (a) Reflection phase of one unit cell for 5 GHz, normal, y-polarized incidence; the white circles depict the series RC settings for the eight unit cells, required for anomalous reflection function. (b) Scattered E_y field pattern for anomalous reflection function showing planar tilted wavefronts to the desired direction. '$+/-$' denote the sign of the values over the two plots. In (a), the value decreases in the clock-wise direction.

The required RC values for the 8 unit cells after optimization are clearly marked in Fig. 3.22(a). With this configuration, the normally incident wave with polarization E_y is reflected to 56.4 degrees without prominent scattering to the specular direction or the $m = -1$

diffraction channel (-56.4 degrees), as the scattered E_y field pattern shows in Fig. 3.22(b). It is important to note that the amplitude of the reflection coefficient is $|S_{21}| = 0.38$, which means that 14.4% of the power is reflected toward the desired direction. Meanwhile, the amplitudes of the reflection coefficients to the other two directions are only $|S_{11}| = |S_{31}| = 0.05$. The absorbed power by the metasurface is given by Abs $\approx 1 - \sum_n |S_{n1}|^2 = 85.1\%$, where n runs through the propagating reflection ports $\{1, 2, 3\}$; an anomalous reflection efficiency of Eff $= |S_{21}|^2 / \sum_n |S_{n1}|^2 = 96.7\%$ is calculated, meaning that 96.7% of the scattered power is channeled to the desired direction. Finally, we emphasize that through the same principle of locally modifying the reflection phase, we can achieve more wavefront manipulation operations, such as retro-reflection and focusing [108, 169].

3.5.3 Polarization Conversion

The proposed unit cell configuration can offer additional opportunities. Organizing patches in groups of four in the unit cell has resulted in a geometrically isotropic unit cell design that can have a polarization independent response. However, the connection of the four patches with electrical loads in the package offers the distinct capability of electrically breaking the four-fold rotational symmetry and thus unlocking the possibility for polarization control [131].

As an example of the opportunities for polarization manipulation, we next demonstrate linear polarization conversion to the orthogonal state, i.e., an incident wave with polarization parallel to the y-axis is transformed to a reflected wave with the electric field along the x-axis. To this end, instead of connecting all four patches to identical loads we appoint different load values to induce the desired asymmetry. Specifically, we utilize the complex loads $\{R_1, C_1\} = \{R_3, C_3\}$ and $\{R_2, C_2\} = \{R_4, C_4\}$ to electrically emulate the geometric structure of a 45-degree tilted cut wire, which has been shown in the literature to result in efficient linear polarization conversion [56]. We use values $\{R_1, C_1\} = \{R_3, C_3\} = \{0.35\ \Omega, 2.6\ \text{pF}\}$ and $\{R_2, C_2\} = \{R_4, C_4\} = \{3.4\ \Omega, 4.8\ \text{pF}\}$ [131], that allow for inducing the desired asymmetry, while being technologically feasible with regards to both resistance and capacitance. Note that for the used values of R and C the reactances dominate over the resistances at the neighborhood of 5 GHz.

The obtained response is depicted in Fig. 3.23, showing perfect polarization conversion to the orthogonal state, i.e., $R_{\text{co}} = 0$, at 5.05 GHz. Since resistance values as high as 3.4Ω have been used for the off-diagonal elements, the absorption is high ($A = 0.61$ at 5.05 GHz), limiting the cross-polarized reflection power coefficient to

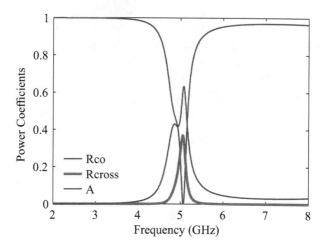

Figure 3.23 Linear polarization conversion to the orthogonal state by electrically breaking the four-fold symmetry using loads $\{R_1, C_1\} = \{R_3, C_3\} = \{0.35 \ \Omega, 2.6 \ \text{pF}\}$ and $\{R_2, C_2\} = \{R_4, C_4\} = \{3.4 \ \Omega, 4.8 \ \text{pF}\}$ to emulate a 45-degree tilted cut wire. Complete polarization conversion to the orthogonal state is observed. The cross-polarized power coefficient is 0.37 and absorption amounts to 0.61. The efficiency and extinction (quantities defined in the text) of the linear polarization conversion process at ~ 5.05 GHz are 0.97 and 27 dB, respectively.

0.37, as shown in Fig. 3.23. The remaining power (2%) is found in transmission (not shown), which is in all cases small but non-zero due to the presence of perforations in the backplane. The performance can be further quantified by the metrics of efficiency (Eff $= R_{\text{cross}}/(1-A)$) and extinction (Ext $= R_{\text{cross}}/R_{\text{co}}$). They describe the percentage of cross-polarized reflection out of the power that does not get absorbed and the completeness of the conversion in reflection, respectively. In Fig. 3.23 their peaks appear at ~ 5.05 GHz and equal 0.97 and 27 dB, respectively.

Increasing the capacitance C_1 inside the range 2.6-3.5 pF, while keeping R_1 constant, we can shift the peak of the cross-polarized reflection to lower frequencies. Specifically, the frequency variation can cover a range of 120 MHz. This frequency blueshift is accompanied by a decrease in the cross-polarized reflected power, since the reactances of the two branches become more comparable; however, the efficiency and extinction remain in all cases very high. Finally, we note that access to lower resistances would reduce the observed absorption and access to a wider span of capacitances would enhance the reactance dissimilarity between the on- and off-diagonal elements, benefiting the polarization conversion performance.

3.5.4 Electromagnetic Characterization Procedures

The aforementioned tunable metasurfaces are fabricated and then the tunable electromagnetic functions are demonstrated experimentally. In this section, we describe the experimental setup and the experimental procedures that will be used to demonstrate the tunable perfect absorption performance, the anomalous reflection and the polarization conversion of the device.

The metasuface sample can be one tile or a combination of several tiles. Each tile consists of 35×35 unit cells and has a size of approximately 5×5 λ. A sample with one tile is shown in Figure 3.24. All the packages within the metasurface (one package per unit cell) are connected and controlled by an external computer and the load configurations can be tuned, in a concomitant or independent way, through the user interface, to modify the incident field from different incidence angles.

3.5.4.1 Experimental Setup

The experiment will be performed in an anechoic chamber and the main parts of the setup, including the sample, the source antenna, and the detecting antenna are schematically shown in Figure 3.25. The tunable metasurface sample is fixed inside the anechoic chamber and controlled by a computer outside. The load values (resistance and capacitance)

Figure 3.24 Multifunctional metasurface including one tile.

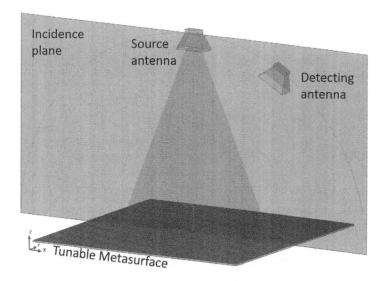

Figure 3.25 Schematic of the experiment setup. The gray plane is the incidence plane. The source and detecting antennas will move along the arc in this plane.

of the packages are tuned by the computer to change the local surface impedances (both resistive part and reactive part) for the selected functionality for different incidence directions. The incident wave is generated by a source horn antenna and the incident angle can be changed by re-positioning the source antenna. For example, it can be moved along the arc in Figure 3.25 to scan an angle span in the $z - x$ incidence plane. Then the scattered wave in different directions can be measured by the detecting antenna. In the schematic of Figure 3.25, only the $z - x$ incidence plane is shown and one can do the same for the $z - y$ incidence plane.

As the horn antenna produces a beam rather than a plane wave, we should consider carefully the setup to decrease the edge effects of the sample. For example, when the sample is with one tile, it is better to use a narrower beam to accommodate the spot size inside the metasurface area and avoid edge diffraction effects.

The beam profile of an antenna is usually characterized by the 3 dB beam width (HPBW), for example, 20° or 56° beam width. On the other hand, the distance between the horn antenna and the sample should be larger than the free-space wavelength to have a plane-wave like wavefront at the metasurface position. This is due to the fact that the sample is designed by simulations with plane-wave excitations. As our tunable metasurface has the size 300 mm by 300 mm and is designed for operation at 5 GHz (i.e., the wavelength about 60 mm), we set the distance between the antenna and the sample to be larger than 720 mm (12 wavelengths). With this consideration, in Figure 3.26 we schematically show the beam profiles in the $z - x$ plane for both normal and oblique incidences (electric field always along the y direction) and for two horn antennas with 3 dB beam widths 20° and 56°.

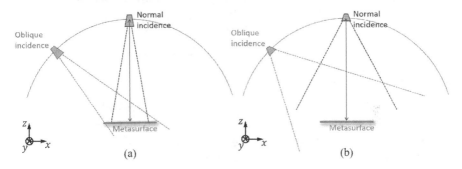

Figure 3.26 Schematic of the beam profiles on the metasurface under normal and oblique incidences for source horn antennas with 3 dB beam widths (a) 20° and (b) 56°.

As we can observe from Figure 3.26(a), for a narrower beam comparing with the wider beam in Figure 3.26(b), most of the energy shines on the metasurface for both normal and oblique incidences. Therefore, we prefer to use a narrower beam. However, there still will be some edge effects for large incidence angles and we will consider them with other means, such as with a larger sample consisting of 2 × 2 tiles or applying the physical optics version of the diffraction theory. For the detecting antenna, we will use the same model as for the source antenna.

3.5.4.2 Demonstration Procedure

Before the actual measurements and demonstrations we need to perform reference measurements with different incidence angles. The reference is taken with a PEC plate with the same size as the sample. The reference PEC plate is on the same position and the scattered fields are measured for different incidence angles. Then, we position the tunable metasurface sample in place and perform the following measurements:

1. Normal incidence: Set the source horn antenna at the normal incidence position, then measure the scattered field for differ-

ent resistance and capacitance settings set through the computer interface. The results can be summarized and the optimal R and C values can be found for the desired function at normal incidence.

2. TE oblique incidences: Set the source horn antenna at oblique incidence angles for TE polarization, for example, $10°$, $20°$, $30°$, etc. (different incidence angles will be demonstrated separately). Then the same scattered field measurement as the normal incidence case will be performed. The optimal R and C values will be found.

3. TM oblique incidences: Similar to the TE oblique incidence.

Following the procedure described above we will certainly demonstrate the tunable electromagnetic functions, similar to that performed in the simulations. On the other hand, the reconfigurable tunable metasurface also enables us to demonstrate its performance in a dynamic way. In this case, the incidence angle is changing with time and the metasurface is reconfigured adaptively to perfectly modify the incident wave from different incidence angles. The dynamic demonstration procedure will be as follows:

1. At arbitrary incidence angle θ_0 (normal incidence, TE oblique incidence or TM oblique incidence) the metasurface is set with the optimal R and C values for achieving the desired function at this incidence angle.

2. The incidence angle is changed to another one, θ_1. Right after the change, we will detect some scattered field from the detecting antenna. Then, the R and C are reconfigured quickly through the computer to the values that support the desired electromagnetic function for the incidence angle θ_1. After the reconfiguration, the detected scattered field disappears and we have perfect absorption.

3. The incidence angle changes continuously in time, and the reconfiguration process (same as the ones in the previous step) follows to dynamically demonstrate the desired tunable function.

Here, we have accepted the fact that we know the incidence angle in advance. This is because the tunable metasurface itself cannot detect the incident wave therefore it cannot adapt to the incidence angle change by itself alone.

3.5.4.3 Spatial Modulation of Load Configuration for Better Performance

In the above discussions, we have assumed that the wavefront of the beam resembles a plane wave for demonstrating the tunable perfect absorber. Therefore, all the packages are tuned in a concomitant way. However, in reality the planar wavefront assumption is not entirely accurate, especially when the horn antenna is close to the sample. The output wavefront from a horn antenna is an arc instead of a vertical plane. Therefore, in the far field, the wavefront of the beam looks like the one emitted from a point source, and at different position of the metasurface the incidence angle is actually different.

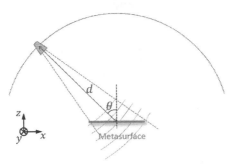

Figure 3.27 Schematic show of the wavefront at the metasurface emitted from a pyramidal horn antenna.

Due to the non-plane wave excitation, the tunable metasurface may not be able to perfectly modify the incident field when the packages are tuned to the same R and C values in a concomitant way. Fortunately, our tunable metasurface allows us to tune the packages in an independent way. In fact, we can calculate the local incidence angle if we know the position of the source antenna. For example, Fig. 3.27 shows the schematic when the incidence angle is θ on the $z-x$ plane. The distance between the source antenna and the metasurface is d. As a result, the incidence angle θ_x at position x is given by

$$\theta_x = \arctan \frac{d \cos \theta}{d \sin \theta + x}. \tag{3.39}$$

Then, the R and C values of the package at position x can be tuned to the values that work for the selected electromagnetic function for incidence angle θ_x. The packages at other positions can be treated in a similar way to enhance the performance. In this way, we expect that the performance will be improved compared to the concomitant tuning case.

3.6 DESIGN OF THE GRAPHENE-BASED PROTOTYPE

Graphene is a two-dimensional material which has received intensive attention in recent years for the applications of photodetectors [54,184], wave modulators [142], and plasmonic devices [71, 172]. The most attractive property of graphene is its tunable band structure which allows to tune its conductance within a large range by applying an external bias voltage [120] or pumping with optical beams [160]. For this reason, graphene layers can replace varactors and varistors used in the microwave range and provide tunability of metasurfaces in terahertz and optical ranges. Similarly with the switch-fabric metasurfaces, the electrical properties of graphene can be globally or locally modified with electrical bias. In the scheme of global tuning, a graphene layer is biased with a uniform voltage. Local modification of graphene conductivity can be achieved by applying different control voltages on separate graphene patches [26] or varying the thickness of supporting dielectric layers [172]. This section is focused on the design of tunable metasurfaces based on globally tuned graphene sheets.

3.6.1 Practical Considerations and Design Constraints

The conductivity of graphene can be described by the Kubo formula [60], which is a function of the Fermi level (controlled by external voltage) E_F and the carrier mobility μ_m:

$$
\begin{aligned}
\sigma_g(\omega, E_F, \mu_m) = & -j\frac{e^2 k_B T}{\pi \hbar^2 (\omega - j\gamma)} \left[\frac{E_F}{k_B T} + 2\ln\left(1 + e^{-\frac{E_F}{k_B T}}\right) \right] \\
& -j\frac{e^2}{4\pi\hbar} \ln\left[\frac{2|E_F| - (\omega - j\gamma)\hbar}{2|E_F| + (\omega - j\gamma)\hbar} \right].
\end{aligned}
\tag{3.40}
$$

Here, $\gamma = ev_F^2/(\mu_m E_F)$ is the scattering rate (v_F is the Fermi velocity), k_B is the Boltzmann constant, e is the elementary charge, $\hbar$ is the reduced Plank constant, and T is the temperature. The first term represents the intraband transitions contribution which is approximated under the condition of $\hbar\omega \ll 2|E_F|$, and the second term refers to the interband contribution. In the terahertz frequency band and at lower frequencies the intraband transitions dominate. Throughout this chapter, we discuss applications in the terahertz band and approximate the surface conductivity of graphene with the first term in (3.40). However, we note that the presented design methodology is applicable in the general case.

From the Kubo formula, we can see that the conductance of graphene can be practically controlled by varying the Fermi level and the carrier mobility. Existing techniques for electrical control of the Fermi level are mature. Using high-capacitance dielectric layers or ion gel, the Fermi level can be tuned within a wide range $(0 - 1 \text{ eV})$. The very critical factor is the control of graphene mobility. The mobility of CVD graphene is largely affected by fabrication environments. In the processes of graphene sample fabrication, inclusions or impurities from external environment significantly decrease the carrier mobility of graphene. The most common achieved mobility value is only around $1000 \text{ cm}^2\text{V}^{-1}\text{s}^{-1}$ [70, 71, 189], corresponding to sheet resistivity of 9000 Ω/sq (at a low Fermi level, $E_\text{F} = 0.1$ eV), which is severely mismatched with the free-space wave impedance. This means that graphene interacts very weakly with the incident light, which also to some extent restricts its tunability. A natural monolayer graphene can absorb only a very small fraction of incident light power (e.g., 2.3% in the optical range).

In this chapter we explain how to use auxiliary metasurface structures to improve absorption in graphene and realize strong tunability of absorption. First of all, we need to understand the exact physical reason for the low absorption in graphene. We start from the simplest absorptive structure, a graphene-Salisbury screen, where graphene is mounted on a dielectric substrate and the transmission is totally blocked by the ground plane, as shown in Fig. 3.28(a). The equivalent circuit of the structure in Fig. 3.28(a) is shown in Fig. 3.28(b).

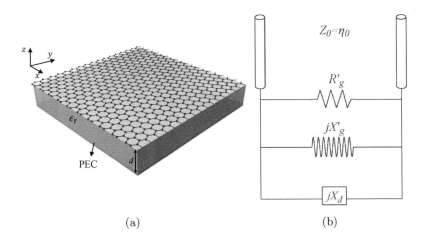

(a) (b)

Figure 3.28 (a) Schematic and (b) circuit model of graphene-based Salisbury screen.

The grounded substrate is modelled as a section of shorted transmission line with the length equal to the thickness of the substrate. Different from the conventional Salisbury screen where the impedance sheet is purely resistive, in a graphene-based structure, since graphene also exhibits inductive response, the graphene layer can be modeled as a series connection of resistance and inductance, $Z_g = 1/\sigma_g = R_g + jX_g$. Alternatively, any series connection can be transformed into parallel form, $Z_g = R'_g \parallel jX'_g$, with the following relations:

$$R'_g = R_g + \frac{X_g^2}{R_g} \tag{3.41}$$

and

$$jX'_g = j\left(X_g + \frac{R_g^2}{X_g}\right). \tag{3.42}$$

This transformation is very useful in the interpretation of weak absorption and for understanding how to improve the absorption, as we will see in the following. Knowing all the circuit parameters, the input admittance of the circuit is the sum of admittances of each shunt component:

$$Y_{in} = \frac{1}{R'_g} + \frac{1}{jX'_g} + \frac{1}{jX_d}. \tag{3.43}$$

Perfect absorption takes place when $Y_{in} = Y_0 = 1/Z_0$, which satisfies the impedance matching condition. Thus, the real and imaginary parts of Y_{in} should satisfy

$$\frac{1}{R'_g} = \frac{1}{Z_0}. \tag{3.44}$$

and

$$\frac{1}{jX'_g} + \frac{1}{jX_d} = 0. \tag{3.45}$$

Equation (3.45) tells that the shunt reactance of the graphene sheet and the input reactance of the grounded substrate should cancel out (forming a high-impedance surface). This condition is usually easy to satisfy because for an arbitrary value of X'_g, it can be easily met by adjusting the thickness or the dielectric constant of the substrate. Equation (3.44) indicates that the shunt resistance of the graphene sheet should match the free-space impedance at the desired frequency of perfect absorption (for normal incidence, $Z_0 = \eta_0 = \sqrt{\mu_0/\epsilon_0}$). Deviation of R'_g from Z_0 determines the absorption level. Assuming that when the condition (3.45) is satisfied, the reflection coefficient can be written as $r = (R'_g - Z_0)/(R'_g + Z_0)$. We see that the reflectance of the structure is completely determined by the shunt resistance of graphene. If R'_g deviates severely from Z_0, the wave interacts weakly with graphene and the absorption is small. Here, we define the logarithm of the normalized shunt resistance as a decoupling factor, $D_e = \log(R'_g/Z_0)$. The decoupling factor measures the degree of impedance mismatch: a wave can be fully absorbed in graphene only if $D_e = 0$; large absolute values of D_e correspond to low absorption.

Next, we study the decoupling mechanisms for graphene with arbitrary mobility and Fermi levels in the terahertz region. Figure 3.29 shows the calculated D_e as a function of graphene mobility and Fermi level at 4 THz. We study the doping range from 0.1 eV to 1 eV, which can be realized with high capacitance ion gel films [66, 72, 141]. For low quality graphene ($\mu_m < 2000$ cm^{2}V^{-1}s^{-1}), we can see that decoupling is strong and absorption is very weak at low Fermi levels. This is caused by the huge sheet resistance of weakly-doped graphene (R'_g is dominated by R_g). Therefore, improving the doping level (decreasing R_g) will somewhat enhance the absorption. This can clearly explain the recent experimental observation of perfect terahertz absorption in graphene [72], where graphene is subjected to strong electrical doping ($E_F = 1$ eV). For high-quality graphene, the impedance mismatch becomes very severe, and it cannot be alleviated by the increase of Fermi level. The decoupling is dominated by the kinetic inductance

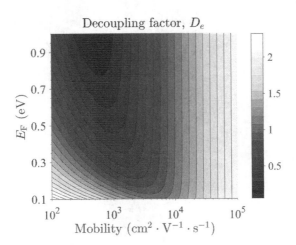

Figure 3.29 Decoupling factor D_e as a function of the carrier mobility and the Fermi level at 4 THz.

of graphene (R'_g is mostly affected by X_g). This theory can success-fully explain the absorption enhancement in graphene plasmonic patterns [165]. In fact, the capacitive interactions between the neighboring graphene cells cancel the natural kinetic inductance of graphene, this way increasing the coupling. We see that at low doping levels, both low- and high-quality graphene sheets are difficult to couple with incident waves, although decoupling is caused by different mechanisms (large sheet resistance or reactance). Hereafter, we aim at perfect absorption in weakly-doped graphene.

It is known that a pristine graphene layer always becomes doped by some residual carriers during the growth and processing of graphene. This initial doping is unintentional, and varies from sample to sample. We take a typical value of this unintentional doping as $n_0 = 7.3 \times 10^{11} \mathrm{cm}^{-2}$, corresponding to the Fermi level of $E_{\mathrm{F}0} = 0.1$ eV [46, 158]. Realizing perfect absorption with such low Fermi levels is more difficult than in heavily-doped graphene, but, beneficially, it may enable the use of graphene in perfect absorbers without external assists (electrical or chemical doping). More importantly, if the matching condition is satisfied in a weakly-doped graphene, it is possible to maximize tunability because one can drive the Fermi level from a very low state to the practical upper limit, which provides a maximum force to break the established matching. Figure 3.30 shows the decoupling factor for $E_\mathrm{F} = 0.1$ eV at different frequencies and for different mobility values. In the high-mobility regime, the inductance-induced decou-

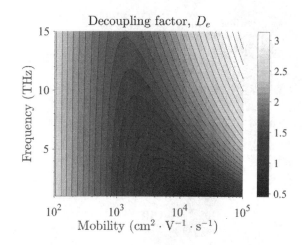

Figure 3.30 Decoupling factor as a function of the carrier mobility and the frequency at $E_{\mathrm{F}0} = 0.1$ eV.

pling becomes evident as the frequency increases (X_g increases fast and becomes much larger than R_g), while for low-quality graphene it is almost frequency-independent (R_g is much larger than X_g), which means that strong decoupling is a universal problem in the whole terahertz range for weakly doped graphene. This indicates that both over-resistive and over-conductive graphene properties restrict absorption. Although these problems are caused by different decoupling mechanisms, in the next section, we demonstrate that both decoupling effects can be eliminated by placing one type of metasurface substrate below graphene, and perfect absorption is always achievable in graphene irrespective of the carrier mobility, only by engineering the dimensions of metasurface patterns.

3.6.2 Graphene Combined with Metallic Patches

Figure 3.31(a) shows the structure of the metasurface-based graphene absorber. Compared to the Salisbury structure (graphene sheet on a metal-backed substrate), we deposit one auxiliary metallic pattern that consists of periodic subwavelength patches on the substrate. The graphene sheet is directly transferred on top of the metallic pattern. The metallic metasurface (formed by a periodic square patch array and the grounded substrate) plays two important roles: creation of a capacitive component to develop a resonance together with inductive graphene (the structure known as a high-impedance surface), as well as reduction of the effective shunt resistance of graphene.

In the presence of metallic patches, the graphene layer is equivalent to a "patterned" fishnet structure (a complementary structure to the array of square patches), as shown in Fig. 3.32(a). This is because in the graphene-metal contact areas graphene is effectively shorted by the highly conductive metal. The tangential electric field on graphene where it is supported by metal is zero due to the boundary condition on metal. Therefore, graphene becomes "invisible" to the incident wave when it contacts metal, and there is no current induced in this part of the graphene layer. The only effective part is the graphene "fishnet" which is not shorted. Assuming that the electric field is oriented along the x-direction, strong capacitive coupling between the patches is induced via y-directed gaps. Therefore, in the effectively "patterned" graphene only y-oriented strips are strongly excited [see the arrows in Fig. 3.32(a)]. Figure 3.32(b) shows the equivalent circuit of one unit cell. The x-polarized electric field is analogous to a voltage source. The graphene sheet is "patterned" into three shunt-connected impedance strips with impedance values Z_x, Z_y, and Z_x. The effective

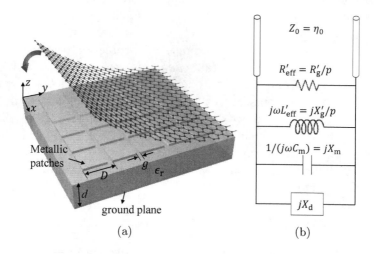

(a) (b)

Figure 3.31 (a) Schematics of the proposed graphene absorber based on metasurface substrate (b) equivalent circuit model of the proposed structure for normal incidence.

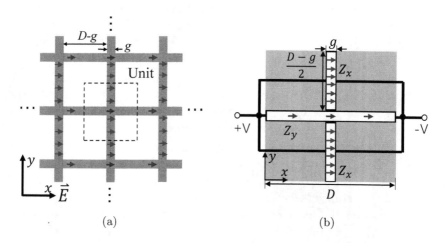

(a) (b)

Figure 3.32 (a) Top view of the "patterned" graphene. The arrows represent weak and strong induced currents in the strips due to capacitive coupling between the patches. The black dashed region represents one unit cell. (b) Circuit analogue of graphene sheet in one unit cell. The gray region is covered by metal, and white strips are graphene.

impedances of the three strips can be calculated as their length-width ratios multiplied by the intrinsic sheet impedance of graphene Z_g (the length refers to the edge along the current flowing direction):

$$Z_x = \frac{2g}{D-g} Z_g, \quad Z_y = \frac{D}{g} Z_g, \tag{3.46}$$

where D and g denote the cell period and the gap width, respectively. Under the condition $D \gg g$ (this assumption holds throughout the chapter), Z_y is very large and it can be viewed as an open circuit, which in turn explains the weak currents in the x-directed strip. Thus, the total effective admittance of graphene can be approximated as

$$\frac{1}{Z_{\text{eff}}} = \frac{2}{Z_x} + \frac{1}{Z_y} \approx \frac{2}{Z_x}. \tag{3.47}$$

From (3.47) we see that the effective impedance of graphene in one cell can be written as $Z_{\text{eff}} = Z_g/p$, where $p = (D - g)/g$ will be called the *scaling factor*. Apparently, if p is large enough, the effective impedance of unshorted graphene will be greatly reduced, which is necessary for the impedance matching with free space.

The equivalent circuit of the proposed structure is shown in Fig. 3.31(b). It is modified from the circuit of the Salisbury screen [see Fig. 3.28(b)] by scaling the graphene impedance down to Z_g/p, as well as adding a parallel capacitive component contributed by the periodically arranged metal patches. In this circuit model, the metal is treated as a perfect conductor. This is a good approximation in the low terahertz band when the metal is gold or silver. The grid impedance of the patch array is capacitive, $jX_m = 1/j\omega C_m$. For a normally incident plane wave, C_m reads

$$C_m = \frac{2\epsilon_{\text{eff}} \epsilon_0 D \ln \left(\csc \frac{\pi}{2(p+1)} \right)}{\pi}, \tag{3.48}$$

where $\epsilon_{\text{eff}} = (\epsilon_r + 1)/2$ [113].

Similarly to the Salisbury screen discussed in the previous section, perfect absorption takes place when the following two conditions are satisfied:

$$\frac{R'_g}{p} = Z_0 \tag{3.49}$$

and

$$\frac{1}{jX'_g/p} + \frac{1}{jX_m} + \frac{1}{jX_d} = 0. \tag{3.50}$$

In what follows, the design procedures for unity absorption at a specified frequency f_0 are introduced. For a given graphene sample with arbitrary mobility μ_{m} and initial Fermi-level E_{F0}, we can obtain the sheet impedance in its shunt form using Eqs. (3.40)-(3.42). Then p is determined from condition (3.49). For a fixed graphene sample and substrate, jX_{d} and jX'_{g} are already known. Thus, D can be uniquely solved to satisfy condition (3.50). However, the obtained D may be much larger than the substrate thickness d. In this case the analytical formula for the grid capacitance (3.48) becomes not accurate due to ignorance of evanescent-mode coupling with the ground plane [168]. In this case, one should use more accurate expressions for C_{m} [168], where the influence of evanescent modes is accounted for. In the following validation examples we setup a relation between the thickness of the substrate and the unit-cell size as $d = aD$, where a is a parameter larger than 0.3. In this case, the analytical formula (3.48) is always valid [42]. With condition (3.50), the patch-array period can then be determined.

After p, D and g have been determined, we can analytically calculate the absorption spectrum. The total input impedance of the structure is

$$Z_{\mathrm{in}} = \frac{Z_{\mathrm{g}} X_{\mathrm{m}} X_{\mathrm{d}}}{p X_{\mathrm{m}} X_{\mathrm{d}} - j Z_{\mathrm{g}}(X_{\mathrm{m}} + X_{\mathrm{d}})}. \tag{3.51}$$

The reflection and absorption coefficients read

$$R = \frac{Z_{\mathrm{in}} - Z_0}{Z_{\mathrm{in}} + Z_0}, \tag{3.52a}$$

$$A = 1 - |R|^2. \tag{3.52b}$$

3.6.3 Frequency-Tunable Perfect Absorber

To design frequency-tunable absorbers, two conditions should be satisfied. The external bias voltage should tune the resonant frequency of the structure. In addition, at every resonant frequency, the input impedance of the structure should match the free-space impedance, $Y_{\mathrm{in}} = 1/Z_0$. Such functionalities can be realized in metasurfaces containing varactors and varistors which can be independently tuned to meet the required resistance and reactance at the desired absorption frequency. Although a single layer of graphene exhibits tunable reactance and resistance, they are not independent, as is obvious from the Kubo formula (3.40).

Here, we show that with a properly structured metasurface substrate, the resonant frequency of the high-impedance surface can be continuously shifted in the terahertz range, while the shunt input impedance of the structure always remains close to Z_0 at the resonance. To realize such functionality, the mobility of graphene is chosen as $\mu_m = 1500$ cm^{2}V^{-1}s^{-1}, and the initial doping is assumed as $E_{F0} = 0.1$ eV. Figure 3.33 shows the analytically calculated absorption coefficient from 1 THz to 14 THz when the Fermi level is raised from the original 0.1 eV to 1 eV. The dimensions of the metasurface substrate are listed in the caption of Fig. 3.33. Apparently, the absorption frequency is blue-shifted from 4 THz to 12 THz, realizing a wide tuning range (about 100%). The peak absorbance at each Fermi level is very close to unity, only with a small degradation around $E_F = 0.2$ eV (92% absorption).

The near perfect tunability in the absorption frequency indicates that the critical coupling between light and graphene is sustainable during the Fermi level modulation, even if graphene's property dramatically changes. It should be pointed out that the mobility of graphene is not strictly limited at $\mu_m = 1500$ cm^{2}V^{-1}s^{-1}. The most effective frequency tuning is achieved with the mobility between $\mu_m = 1000$ cm^{2}V^{-1}s^{-1} and $\mu_m = 2000$ cm^{2}V^{-1}s^{-1}. This mobility range is also quite reasonable and can be obtained using CVD-grown graphene [70, 71, 189, 190].

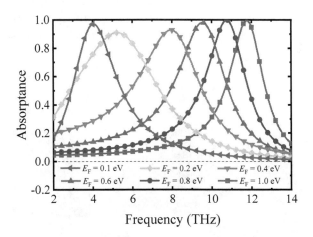

Figure 3.33 Simulated absorption intensity in terms of the frequency and doping level when varying the Fermi level from 0.1 eV to 1 eV. The graphene mobility is set to $\mu_m = 1500$ cm^{2}V^{-1}s^{-1} ($\tau = 15$ fs at $E_F = 0.1$ eV). The results are obtained by choosing the metasurface parameters as, $D = 17$ μm, $p = 15.9$, $d = 0.5D$, and $\epsilon_r = 2.8$.

3.6.4 Switchable Absorber

Let us assume that the mobility of graphene is extremely low ($\mu_{\mathrm{m}} = 200 \text{ cm}^2\text{V}^{-1}\text{s}^{-1}$) and study the effect of electrical tuning on the absorption level. Following the presented design procedure, the structural parameters are determined as $D = 7$ μm, $p = 137$ and $a = 0.5$ if the perfect absorption is expected at 4 THz for $E_{\mathrm{F0}} = 0.1$ eV.

We note that in order to reduce the high resistance of such low-quality graphene, a narrow gap ($p = 136$ and $g = 50$ nm) between the metallic patches is required in this case. This significantly increases fabrication difficulties. The situation will become even worse if the operating frequency is higher (D becomes smaller). On the other hand, the tiny width of the metallic slots is not desirable, since the Fermi level pinning effect in graphene is not negligible. This practical issue is caused by the transfer of electrons in metal to the graphene sheet via the metal-graphene contact. The transferred charges result in an inhomogeneous distribution of the Fermi level in graphene [70, 77].

In order to avoid these problems, we propose an alternative metallic pattern with meandered metal channels instead of straight ones (formed by the patch array), as shown in Fig. 3.34. The y-oriented metal strips (called "fingers") intersect with each other to form a strong capacitance. The enhanced electric flux between the fingers dramatically increases the absorption in graphene. From the perspective of impedance matching, the meandered structure increases the length-width ratio of graphene, so that we can effectively decrease the shunt

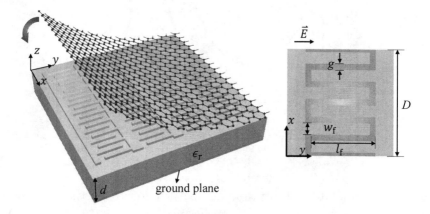

Figure 3.34 Schematic of the metasurface pattern with meandered slots (one square unit). The electric field is along the "finger" direction. This structure is appropriate only for y-polarized incident wave.

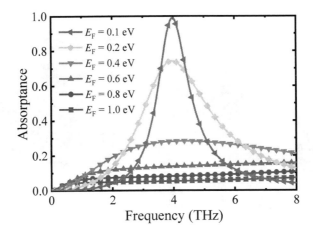

Figure 3.35 Absorption spectrum of the proposed meandered structure. The perfect absorption is optimized at 4 THz when $E_{F0} = 0.1$ eV and $\mu_m = 200$ cm^{2}V^{-1}s^{-1} ($\tau = 2$ fs). The substrate is chosen as $\epsilon_r = 2.8$ and $d = 2.1$ μm.

resistance of the graphene sheet while keeping the gap size manageable in nanofabrication. The required relations between the finger length, the unit and gap sizes can be approximated as $(Nl_f + D)/g = p$, where N is the numbers of fingers in one square unit cell. Using graphene of the same quality ($\mu_m = 200$ cm^{2}V^{-1}s^{-1}), perfect absorption is achieved when $D = 5$ μm, $l_f = 3.6$ μm, $g = 250$ nm and $N = 8$. Perfect absorbance is shown in Fig. 3.35 (tallest curve). When the Fermi level is tuned from 0.1 eV to 1 eV, the absorption level sharply decreases from unity to 5%. The device acts as a perfect absorber in its natural state ($E_{F0} = 0.1$ eV), while in the biased state it is a nearly perfect reflector.

3.6.5 All-Angle Perfect Absorber

Resonant metasurface perfect absorbers operate only at one specific incident angle. For illuminations at other angles the absorbance essentially reduces. This is because at other illumination angles both the characteristic impedance of waves in outside space and the input impedance of resonant structures change, resulting in impedance mismatch. For example, the impedance of TM polarized light drops as the incident angle increases as $Z_0(\theta) = \eta_0 \cos \theta$.

For realization of perfect absorption at every incident angle, two conditions are required, one is to obtain a tunable resistive layer which

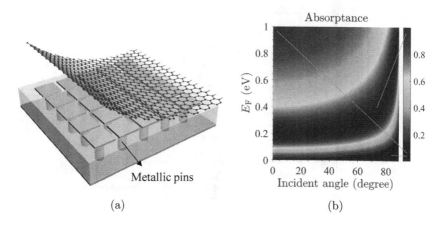

Figure 3.36 (a) Proposed structure for angle-tunable perfect absorbers and (b) absorptance in terms of the incidence angles and Fermi levels at 220 GHz.

can adapt to the change of the wave impedance, and the other is creating a resonant structure insensitive to the incident angle. The first requirement can be realized below the terahertz range by employing graphene sheets. In this frequency range, the sheet resistance of graphene can be effectively tuned while its sheet inductance is negligibly small compared to its resistance. The second condition can be satisfied by embedding periodic metallic pins in the dielectric of Fig. 3.31(a), forming a mushroom structure, as shown in Fig. 3.36(a). Under oblique TM-illumination, the metallic pins enforce the electric and magnetic fields in the dielectric orthogonal to the pins [186]. For this reason, at different incident angles, the characteristic impedance in the substrate is always the same as that of the TEM wave at normal incidence, and therefore the resonance frequency of the structure does not change [166].

As an example, we target realization of perfect absorption at 220 GHz. We assume that the mobility of graphene is $\mu_m = 1000$ cm^{2}V^{-1}s^{-1} [70, 81, 141] and initially set the Fermi level as $E_F = 0.18$ eV, which corresponds to sheet resistivity of $Z_s = 2640$ Ohm/sq.

When the graphene sheet is placed on an array of metallic patches with the structural parameter $p = 7$, the effective shunt resistance of graphene reduces to $Z_s/p = 377$ Ohm, corresponding to the characteristic impedance of normally incident plane waves. Therefore, perfect absorption occurs at the normal incidence. The frequency of perfect absorption (the resonant frequency) is only determined by the supporting metasurface. If we increase the Fermi level of graphene, the effective shunt resistance can continuously match the characteristic impedance at different incident angles, without changing the resonant frequency. Figure 3.36(b) shows that the device can realize perfect absorption continuously from $\theta = 0°$ to $\theta = 88°$ by changing the Fermi level of graphene from 0.18 eV to 1 eV.

3.7 SUMMARY

In this chapter, we presented basic information on metasurfaces, including electromagnetic modeling, tunability mechanisms, gave examples of tunable metasurfaces, and discussed their advantages over other planar materials in realizing various reconfigurable functionalities. Toward the implementation of software-defined metasurfaces, we presented and discussed the constraints and practical considerations in designing such reconfigurable metasurfaces, and realized a switch-fabric based design. Several reconfigurable functionalities, such as tunable perfect absorption, anomalous reflection, and polarization conversion have been realized and discussed. Moreover, we also discussed several graphene-based tunable metasurface designs, for achieving various tunable absorption functions.

ACKNOWLEDGMENTS

This work was funded by the European Union's Horizon 2020 research and innovation program, Future and Emerging Technologies (FET Open) under grant agreement No 736876. The authors acknowledge the fruitful discussions with Julius Georgiou, Marco A. Antoniades, Kypros M. Kossifos, Loukas Petrou, Giorgos Varnava, and Petros Karousios from the University of Cyprus, Dionysios Manessis and Manuel Seckel from the Fraunhofer Institute for Reliability and Microintegration (IZM), and George Deligeorgis from the Foundation for Research and Technology Hellas (FORTH).

CHAPTER 4

Designing the Internet-of-Materials Interaction Software

Ageliki Tsioliaridou, George Pyrialakos, Alexandros Pitilakis, Sotiris Ioannidis, Nikolaos Kantartzis

Foundation for Research and Technology Hellas, 70013, Heraklion, Crete, Greece

Christos Liaskos

Foundation for Research and Technology Hellas, 71110, Heraklion, Crete

CONTENTS

4.1	Software Design Considerations		78
4.2	The Internet-of-Materials Software Architecture		79
4.3	The Internet-of-Materials Application Programming Interface		84
	4.3.1	General Use Case Diagram	84
	4.3.2	The Database Diagram	88
		4.3.2.1 Table "DoA"	92
		4.3.2.2 Table "Polarities"	92
		4.3.2.3 Table "SwitchStates"	93
		4.3.2.4 Table "Physical Setup"	94
		4.3.2.5 Table "Function Electromagnetic Profiles"	95
	4.3.3	The Class Diagram	96
4.4	A Novel Software Class: The Electromagnetic Compiler		97
	4.4.1	A Qualitative View of the Compiling Process	97
	4.4.2	Metasurface Functions	100
	4.4.3	Formal Definition of a Metasurface Configuration	101
	4.4.4	Definition of Fitness Function	103

	4.4.5	Methods	104
4.5		Theoretical Foundations of the Electromagnetic Compiler	105
	4.5.1	Definitions	106
	4.5.2	Floquet (Unit-Cell) Analysis	107
	4.5.3	ABSORB Functionality	111
	4.5.4	REFLECT Functionality	113
	4.5.5	POLARIZE Functionality	113
	4.5.6	STEER Functionality	114
	4.5.7	SPLIT Functionality	117
	4.5.8	Far-Field Scattering/Radiation Pattern	118
	4.5.9	Formal Definition	118
	4.5.10	Semi-Analytical Calculation	119
	4.5.11	Polarization	121
	4.5.12	Scattered Power in a Lobe (Solid Angle Cone)	123
	4.5.13	Fitness Functions per Functionality	124
		4.5.13.1 ABSORB Functionality	124
		4.5.13.2 STEER Functionality	124
		4.5.13.3 REFLECT Functionality	127
		4.5.13.4 SPLIT Functionality	127
		4.5.13.5 POLARIZE Functionality	127
		4.5.13.6 FOCUS and COLLIMATE Functionalities	128
		4.5.13.7 SCATTER Functionality	129
		4.5.13.8 ARBITRARY Functionality	130
	4.5.14	The Configuration Optimization Process	130
4.6		Software Aspects of the Electromagnetic Compiler	132
	4.6.1	General Use Cases	132
	4.6.2	Validating the Compilation Outcomes with Measurements	138
4.7		Conclusion	140

4.1 SOFTWARE DESIGN CONSIDERATIONS

A major novelty of HyperSurfaces is that they provide a software model to interact with the physical capabilities offered by metasurfaces in general. As such, the context of this chapter is to study the exposed HyperSurface application programming interface (API), its transformation (electromagnetic compilation) to controller directives and their relay within the controller network [99, 101]. This

API and middleware, clearly specified in a standard software modeling approach, will expose the HyperSurface functionality in the form of parametric virtual functions, each corresponding to a metasurface capability (e.g., impinging wave steering, absorption, polarization and any non-linear response in general. Such functionalities are useful in a wide variety of setups in next generation wireless communications [92,93,95,100,102,103,138,161,170], described later in Chapter 9.

Following these considerations, the current chapter will introduce the physical system properties to be controlled, i.e., the electromagnetic behavior of a metasurface, describe how this control can be enforced via electronic means, and proceed to detail the software design that will be employed by an end-user for setting and obtaining the desired type of electromagnetic behavior.

4.2 THE INTERNET-OF-MATERIALS SOFTWARE ARCHITECTURE

We proceed to make a coarse, computer science-based modeling of the HyperSurface operation. To better understand the programming entities involved, we first provide an overview of the physical setup.

An external programming entity ("Caller"), which could be a regular computer, obtains an implementation of the HyperSurface API. It then executes a HyperSurface command. The command is translated to data packets that are sent to the HyperSurface Gateway using a given communication protocol. The Gateway diffuses the information within the controller network of the HyperSurface using an inter-controller communication protocol. The controllers receive the information pertaining to them and set the states of the switch elements accordingly. The collective end-state of the switch elements is then matched to the intended electromagnetic function of the HyperSurface.

Based on this coarse description, we identify the following conceptual entities involved in the HyperSurface API workflow:

1. **The Caller or User.** The entity that calls the HyperSurface API.

2. **The User Service.** A user-side daemon that receives and handles interrupts, e.g., messages spontaneously generated by tiles.

3. **The Callback.** A single function of the HyperSurface API.

4. **The Configuration.** A data entry containing the information for mapping a Callback to a specific set of Switch Element states.

5. **The Configuration Database (DB).** A database containing several discrete configurations.

6. **The Configuration Resolver.** The software that resolves a Callback to a given Configuration.

7. **The Switch Element.** A switch element of the HyperSurface. It can be set to any state within a set of possible states.

8. **The Switch State.** The state of a switch element. It can comprise several impedance values, i.e., discrete combinations of resistance and capacitance.

9. **The gateway external communication protocol.** It is the protocol that transfers data between the HyperSurface and the Caller.

10. **The Gateway.** The HyperSurface Gateway hardware, as described in the proposal.

11. **The HyperSurface controller communication protocol.** The protocol that transfers data between the gateway and the switch controllers, as well as between the controllers themselves.

12. **The HyperSurface Tile.** The complete, assembled HyperSurface unit.

We now define the tentative form of the API callbacks. As briefly mentioned earlier, these callbacks have the following general form:

outcome ← callback(actiontype, parameters)

The **actiontype** is an identifier denoting the intended function, e.g.,:

1. (a) STEER,
   (b) ABSORB,
   (c) POLARIZE,
   (d) FILTER.

Each action type is associated with a set of **parameters**. For instance,

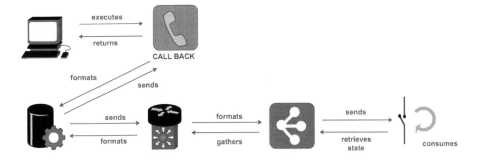

Figure 4.1 The HyperSurface API workflow.

1. (a) STEER commands require:

i) an incident wave direction, I,
ii) an intended reflection direction, O,
iii) the applicable wave frequency, F.

2. (a) ABSORB commands require no O parameter.

An important point is that the HyperSurface API serves as a strong layer of abstraction that hides the internal complexity of the HyperSurface. It offers user-friendly and general-purpose access to metasurface functions without requiring knowledge of the underlying hardware and Physics.

We proceed to illustrate the API Callback process for setting the state of a switch element as shown in Fig. 4.1. The Caller executes an electromagnetic Function deployment Callback function, which in turn invokes the Configuration resolver. The resolver queries the Configuration DB and returns one or more Configurations that are combined into the configuration that best matches the intended electromagnetic function. The Configuration is conveyed to the Gateway using the corresponding protocol. The Gateway re-formats the received configuration and executes actions to inform (and thereby set) each separate switch element accordingly. The HyperSurface controller protocol is employed towards this end. Finally, each switch element is set to its intended value, thereby "consuming" the corresponding part of the configuration.

The configuration resolver may need to combine several configurations to produce the intended one. The reason for this can be explained using the illustration of Fig. 4.2. This figure illustrates a HyperSurface illuminated by an electromagnetic wave incoming from a vertical

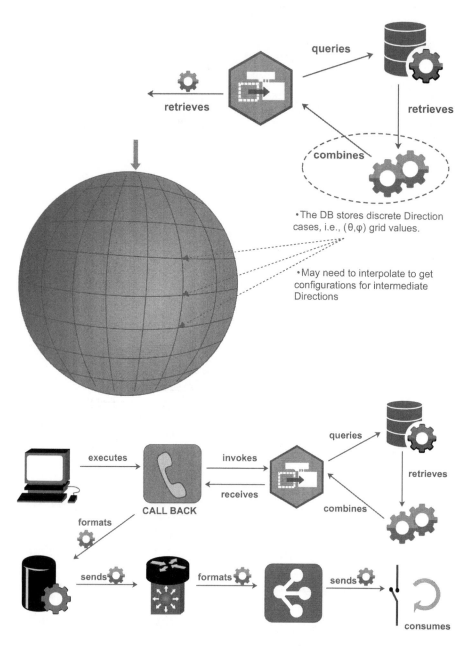

• The DB stores discrete Direction cases, i.e., (θ,φ) grid values.

• May need to interpolate to get configurations for intermediate Directions

Figure 4.2 A singe API callback combining several HyperSurface configurations to achieve a complex functionality.

impinging direction (angles $\theta=0$, $\phi=0$, using a spherical coordinate system relative to the middle point of the HyperSurface). Let us assume that the required function is of type ABSORB. The DB will contain several configurations corresponding to combinations of (ϕ, θ) angles, as shown by the spherical mesh. However, these combinations are discrete. Therefore, the Configuration resolver may need to return configurations for combinations that do not exist in the DB.

Possible directions to be explored for the combination of several configurations include:

1. The selection of the best matching configuration out of the intended ones.

2. A linear interpolation of existing configurations.

3. A non-linear interpolation approach, e.g., using a trained neural network.

We note that apart from setting the state of the switch elements, the API callbacks include functions for getting the state of the elements. We refer to these functions as monitor callbacks due to their immediate usage to monitor the state of the HyperSurface. In Fig (4.3), the Caller executes a monitor Callback to get the state of a switch element. Notice that this callback type requires no interaction with the configuration resolver or the configuration DB. This monitoring technique is called polling since switch elements are queried for their state.

Additionally, depending on the hardware capabilities of the controller nodes, the nodes themselves may trigger a monitoring event, such as a local malfunction. This approach, called reporting, is illustrated in Fig. 4.3.

Notice that this approach requires a service to be active at the side of the Caller that is ready to receive incoming reports from the controllers. The Callback API specification will include a description of such services, if they are supported by the hardware capabilities.

In both polling and reporting monitoring approaches, time-outs or other transmission failures may occur for a variety of reasons (malfunction, interference, and power supply issue). The handling of these events will also be described in the Callback APIs by taking into account the behavior and capabilities of the involved hardware (Gateway and Controller nodes).

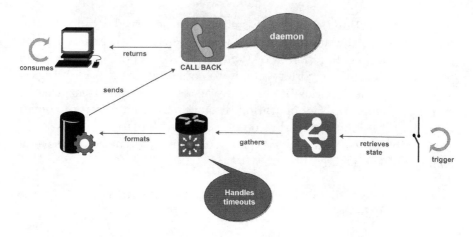

Figure 4.3 The workflow of a callback to monitor the state of a Hyper-Surface element.

4.3 THE INTERNET-OF-MATERIALS APPLICATION PROGRAM-MING INTERFACE

This section details the operation of the API using principles of the Unified Modeling Language. We begin by presenting the use-case diagram which provides a high-level view of the API functionalities. Then, we proceed to describe a high-level Database diagram which provides an insight of the software representation of the involved entities. Then, a tentative class diagram provides the overview of the HyperSurface software modeling. Finally, various diagrams provide the planned functionality of several class member functions.

4.3.1 General Use Case Diagram

The use case diagram of the API is given in Fig. 4.4. The involved entities are the following:

1. The **User** represents the entity than can initiate API callbacks. In this sense, the User may represent a third-party application that includes the API callbacks in its source code or a Graphical User Interface (GUI) that receives commands from a user (e.g., button click events) and subsequently triggers the corresponding API callback.

2. The **Interrupt handling service** is a persistent daemon (service) that receives and dispatches events originating from the

HyperSurface hardware. For instance, a HyperSurface tile may report data obtained from a gateway sensor in a periodic fashion or upon an environmental trigger without the intervention of the user. We note that implementation-wise, this entity may not be implemented as a service at the operating system level. It can instead be implemented as a persistent execution thread that runs in parallel to the user callback execution thread. In this sense, it can be exemplary implemented as a thread within the aforementioned third-party application.

3. The **HyperSurface Gateway** represents the master electronic controller functionalities incorporated in the gateway. It stands between the intra-tile controllers and the user callbacks and acts as a representative of the tile to the external world. Sensing capabilities may be incorporated to the gateway as well.

4. The **Configuration Database (DB)** is the persistent storage point of the API. It contains the allowed HyperSurface electromagnetic functions and their corresponding switch states per tile. Additionally, it contains the allowed HyperSurface subcomponents states, parameters and persistent data required for the API initialization.

The User, the Interrupt Service and the Database are located to common personal computing devices, such as desktops, laptops, or smartphones. They can be located within the same physical device or at completely separate (but networked) ones, without further limitation. The HyperSurface Gateway is bound to each tile.

We proceed to detail the use cases mentioned in Fig. 4.4.

The user can initialize the API prior to any other callback. The callbacks generally refer to i) retrieving the state of a tile in the sense of retrieving the electromagnetic function that is presently deployed over it; and ii) setting it to a required electromagnetic function.

The **Environment initialization** is called once per active user. It performs the following tasks:

1. It detects the *tiles that are presently* active and connected (discoverable) to the user by broadcasting a corresponding network message. The tiles that reply to the message report their identifier within a time threshold. The tiles that do so are logged as active in a browsable, user-space software structure.

2. It *enumerates the supported tiles* by contacting the Configuration DB service thereby obtaining the list of unique tile identifiers.

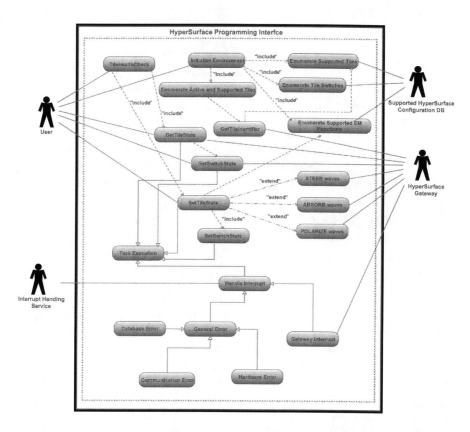

Figure 4.4 Use-case diagram of the HyperSurface API.

3. It *validates the support for the presently active tiles* by checking if their identifier exists in the stored list of supported tiles.

4. For each supported tile, it contacts the DB to retrieve *the switches that comprise the tile* (including their location on the tile and their allowed states), as well as *the electromagnetic functions* (including input and output parameters) *supported by the tile.*

Using the information returned by a successful initialization, a user may perform any of the following tasks:

1. *Set the tile state* to the required electromagnetic function with proper input and output parameters. This is accomplished by setting the state of the tile's switches to the values obtained during the initialization. Three types of electromagnetic functions are planned at this point:

   (a) STEER functions, which express the redirection of waves impinging on a tile from a given direction of arrival (input) to another (output).

   (b) ABSORB functions, which express the maximum achievable absorption (e.g., 99% or more) of an electromagnetic wave impinging on the tile from a given direction of arrival (input).

   (c) POLARIZE functions, which express the alteration of the wave polarization from an initial state (input: initial polarization and direction of arrival) to another (output: initial polarization and direction of arrival).

2. *Get the tile state*, i.e., deduce the currently deployed electromagnetic function on a given tile. This can be accomplished in two ways, depending on the gateway hardware capabilities:

   (a) The currently deployed electromagnetic function identifier may be persistently stored at the tile gateway. In this case, getting the deployed electromagnetic function is implemented by querying the gateway for this identifier.

   (b) The most robust approach is to query the gateway for the current state of each contained switch. The returned states are compared to the allowed tile configurations retrieved from the DB during initialization thereby obtaining the corresponding electromagnetic function identifier.

3. *Check the "Health" of a tile*, i.e., deduce the working condition of the gateway and the tile switches. This is done by requesting the setting of a default state within a time limit. The state is then retrieved back and cross-checked for equality to the default. This task naturally incorporates getting and setting the state of the tile switches.

4. *Getting the state of one or more tile switches* is also allowed to the user for potential GUI visualization and experimental purposes. However, the user is prohibited from setting a tile switch state, as this may break the overall correspondence of the switch states to a valid electromagnetic function.

The interrupt handling service is responsible for dispatching the following events:

1. *Database Errors*, corresponding to invalid data queries or null results.

2. *Communication Errors*, corresponding to network-related errors, such as timeouts.

3. *Hardware Errors*, corresponding to the detection of failing hardware components of the tile.

4. *Gateway Interrupts*, which are events raised spontaneously by the tile gateway without user intervention. These can be simple notifications (e.g., a warning regarding the state of the tile/switches) or they may carry useful data, such as periodically or event-triggered reports on quantities obtained from the sensor at the gateway side.

The aforementioned errors and interrupts can be generalized as Generic Interrupts. Moreover, for the sake of presentation, the main API functionalities (GetTileState, Get/SetSwitchState, Get/SetTileState, InterruptHandling, HealthCheck) are generalized and referred to as *Tasks* that are processed by the API workflow.

4.3.2 The Database Diagram

We proceed to detail the Configuration DB structure in terms of SQL Tables, Relations and Restrictions. While the Configuration DB describes the persistence storage of the HyperSurface API, it also provides a natural view of the software-level modeling of the electromagnetic HyperSurface aspects.

We begin by listing and giving a coarse description of the DB tables. A detailed description per table follows.

The following tables are stubs, i.e., they can be considered as simple data holders, referenced by non-stub tables via foreign-key fields:

1. **Table DoAs.** It stores in rows all considered EM wave directions of arrival (DoAs), over all tiles, which can serve either as input or as output parameters to electromagnetic functions.

2. **Table Frequencies.** It stores in rows the considered electromagnetic function frequencies, serving as input parameters to electromagnetic functions.

3. **Table FunctionTypes.** It stores string identifiers of the considered electromagnetic function types, e.g., "STEER", "ABSORB" and "POLARIZE".

4. **Table Polarities.** It stores the considered polarization effects of electromagnetic functions on waves impinging on the tile.

5. **Table Sourceloc.** It stores the considered source location coordinates for all point sources.

6. **Table Splitdirection.** It stores the considered direction and total number of beams for the beam splitting operation.

7. **Table Switches.** It stores the identifiers of all switches across all tiles.

8. **Table SwitchStates.** It stores all allowed switch states across all controllers and tiles.

9. **Table Unit Cell.** It stores the identifiers of all unit cells that can be handled (i.e., supported) based on the current data in the DB.

10. **Table Simtemp.** It stores results from all simulations performed during an optimization process.

The following tables are complex, relying on foreign keys to the stub tables.

1. **Table Tiles.** It holds the identifiers of all Tiles that can be handled (i.e., supported) based on the current data in the DB.

2. **Table Variables.** It holds the intra-cell parameters associated with all switch elements stored in the *Switches* table.

3. **Table ParameterizedFunctions.** It holds all considered electromagnetic functions (i.e., function types and input-output parameter combinations) per tile.

4. **Table FunctionelectromagneticProfile.** It holds the overall electromagnetic behavior of each parameterized function by detailing reflection/refraction directions and losses and polarity effects, even for unintended inputs (i.e., even when a tile is illuminated by a direction of arrival different than the one considered by the presently deployed parameterized electromagnetic function).

5. **Table FunctiontoConfig.** It holds the switch state configurations that correspond to each entry (row) of the *ParameterizedFunctions* Table. In other words, it provides the compilation data from the intended electromagnetic function to the corresponding, specific tile switch states.

6. **Table PhysicalSetup.** It holds the structural information of each supported tile. Specifically, it holds the switch positions (indexes) per tile, as well as the allowed states per switch.

The database diagram showing the relations among the tables is shown in Fig. 4.5. The *ParameterizedFunctions* Table refers via Foreign Keys:

1. To the stub tables, which serve as inputs/outputs to electromagnetic functions. Notice that DoAs is referred to twice, once as function input, and once as function output.

2. To table Tiles, which provides the context (i.e., the supporting tile) of the electromagnetic function.

3. To table Functionelectromagneticprofile, which describes the electromagnetic behavior of a tile when an electromagnetic function is deployed in it.

The *Functionelectromagneticprofile* Table refers via Foreign Keys:

1. To stub table DoAs twice, once for any impinging DoA and once for any resulting reflection/refraction direction.

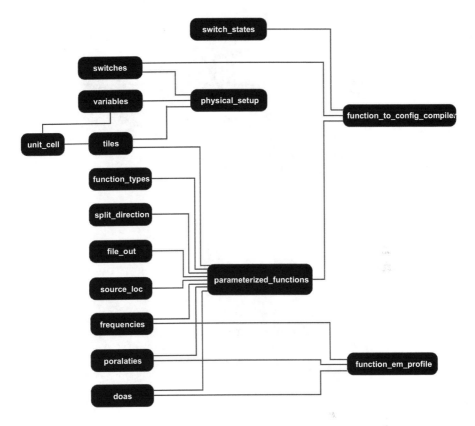

Figure 4.5 The HyperSurface Database diagram.

2. To stub table Frequencies so as to indicate the tile behavior at a given impinging wave frequency (intended by the profile or not).

3. To stub table Polarities so as to indicate the effects on the polarization of the impinging wave per resulting reflection/refraction DoA.

The *FunctionToConfig* table refers via Foreign Keys:

1. To table ParameterizedFunctions so as to indicate the switch states corresponding to each electromagnetic function.

2. To stub tables Switch and SwitchStates so as to indicate the intended (valid) state of each unique switch.

Finally, the PhysicalSetup table refers via Foreign Keys to stub tables Tiles (switch owner), Switches (so as to associate persistent indexes) and to SwitchStates (so as to associate valid states).

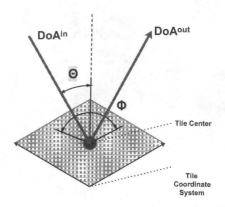

Figure 4.6 Physical meaning of the DoA fields.

We proceed to visualize the data described by a few core data tables.

4.3.2.1   Table "DoA"

The Table DoAs is a stub-table containing all the Tile-Impinging Wave Directions that are considered by any tile and any EM function.

It can be perceived as having three columns:

1. The "ID", which is the primary key,

2. The (ϕ) and (θ) columns, which serve as spherical coordinates defining a wave direction impinging on the tile center, as shown in Fig. 4.6. A tile corner (e.g., bottom-left) is selected by convention to serve as the coordinate system origin.

4.3.2.2   Table "Polarities"

The Table Polarities is a stub-table containing all the polarities that are considered by any tile, and by any electromagnetic function, as either input or output parameters. It can be modeled as having three columns:

1. A unique primary key.

2. The "Q" parameter, which is the normalized factor between the two field vectors in the transverse plain.

3. The "ω" phase difference of the two field vectors,

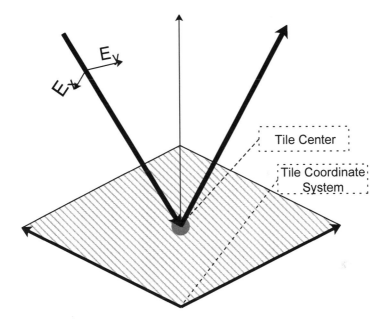

Tile Center

Tile Coordinate
System

Figure 4.7 Modeling the electromagnetic wave polarization.

where the Q factor and phase difference ω are defined through

$$Ex = \sqrt{\tfrac{1+Q}{2}}, \quad Ey = \sqrt{\tfrac{1-Q}{2}}e^{i\omega}, \quad -1 < Q < 1 \ ,$$

essentially describing the direction of the Jones vector for the wave impinging upon the HyperSurface, as shown in Fig. 4.7.

4.3.2.3 Table "SwitchStates"

The Table SwitchStates is a stub-table containing all the switch states that are considered by any switch in any tile. It too can be modeled having three columns:

1. A unique primary key.

2. A varistor value, "R" field.

3. A varactor value, "X" field.

"R" and "X" represent the real and imaginary parts of the total impedance of a Switch element. "R", resistance, is a real positive number. "X", reactance, is also real, but can take both negative and positive values, when the impedance exhibits capacitance and inductance

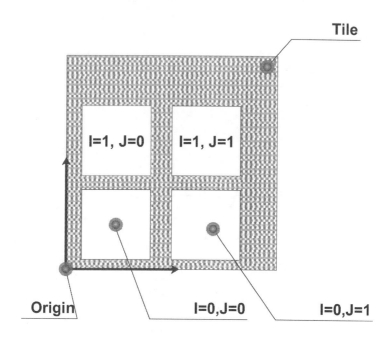

Figure 4.8 The indexed position of the unit cells in the tile. Each unit cell can contain several switch elements.

respectively. This modeling approach is quite generic, since it can freely represent any equivalent circuit topology (with any combination of resistance, capacitance and impedance elements).

4.3.2.4 Table "Physical Setup"

The Table PhysicalSetup is a complex table describing the structure of all the supported tiles based on the data present in the database. Several rows describe the structure of a single tile which is identified by its ID. It can be modeled as having six columns:

1. The primary key.

2. The "TILEID", which is the identifier of a physical tile. ("TILEID" is a foreign key to table "Tiles".)

3. The "SWITCHID" which is the identifier of a switch contained in the tile with ID "TILEID". ("SWITCH ID" is a foreign key to table "Switches".)

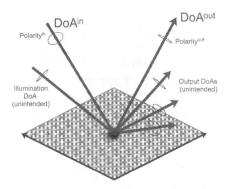

Figure 4.9 Schematic of an example electromagnetic behavior of a tile configured for a given, parameterized electromagnetic function.

4. The fields "CELLI" and "CELLJ" which describe the indexed position of the unit cell containing the switch with id "ID", as shown in Fig. 4.8.

5. The "VARIABLEID" points to the intra unit cell position defined in the respective entry of the "Variables" table. ("VARIABLEID" is a foreign key to table "Variables".)

Figure 4.9 shows an example of how multiple polarization and steering functionalities can be combined for a complex behavior.

4.3.2.5   Table "Function Electromagnetic Profiles"

The Table *Functionelectromagnetic*P*rofile* is a complex table describing the complete electromagnetic behavior of all the supported tiles based on the data present in the database. Several rows describe the structure of a single tile which is identified by its ID.
It has seven columns:

1. The "ID" which is the primary key.

2. The "PARAMETERIZEDFUNCTIONID" which is the identifier of the electromagnetic function that is profiled. ("PARAMETER-IZEDFUNCTIONID" is a foreign key to the table "Parameter-izedFunctions".)

3. The "ILLUMINATIONFREQUENCYID" which is the identifier of a frequency of a wave impinging on the tile. ("ILLUMI-NATIONFREQUENCYID" is a foreign key to the table "Frequency".) This allows for the description of tile behaviors at

frequencies other than the ones intended by the active electromagnetic function.

4. The "ILLUMINATIONDOAID" which is the identifier of a direction of arrival of a wave impinging on the tile. ("ILLUMINATIONDOAID" is a foreign key to the table "DoAs".) The table contains both the DoA expected by the parameterized function, as well as DoAs that are unintended. This describes the electromagnetic behavior of a tile that has been configured for any given function (including cases with a point source input where DoAin is Null), but is unintentionally illuminated by another DoA.

5. The "OUTDoAID" which points to a DoA expressing a major reflection/refraction of the illuminating wave. ("OUTDoAID" is a foreign key to the table "SwitchStates"). Multiple reflection/refraction DoAs are allowed to correspond to one intended or unintended illumination DoA.

6. The "OUTPOLARITYID" which points to a polarity alteration. ("OUTPOLARITYID" is a foreign key to the table "Polarities"). One polarity alternation is allowed per OUTDoAID.

7. The "LOSSDB" field which describes the attenuation loss in decibels (dB) with regard to the power of the illuminating wave. One loss value per OUTDoAID is allowed.

Multiple entries (rows) describe one electromagnetic function profile. This modeling implies only one impinging wave (planar) per function. This follows the common assumptions for metasurfaces [89]. However, multiple input directions can be handled by transforming the "ILLUMINATIONDOAID" to a set of DoA identifiers, rather than just one DoA identifier.

4.3.3   The Class Diagram

In the preceding sub-section we presented the HyperSurface software model by outlining the prospective use cases and the HyperSurface data model. To that end, the persistence model, i.e., the Database, was employed as a practical approach for conveying the overall model. The HyperSurface API is complemented by the Class Diagram shown in Fig. 4.10, i.e., the software process schematics that implement the use cases and interact with the data model.

The Class Diagram is of course closely related to the Database Diagram described in the preceding section. The classes describe objects that are initialized based on the persistent data stored in the Database. Thus, many class objects can be seen as structured data with functionalities.

The API classes can be classified as:

1. Read-only data representers, with no additional functionality (Polarity, Frequency, ParameterizedFunction, FunctionelectromagneticProfile, DirectionOfArrival, SwitchState, SwitchPosition). These simply implement the Interface DBDataHolder.

2. Representers of physical objects, comprising data and functionalities (Tile, Switch). Their data representation aspect is obtained by implementing the DBDataHolder interface as well.

3. Workflow Handlers (Environment, InterruptHandler), which are singleton classes that initialize the API and handle the execution of tasks described in Section 6.2.

4. Interrupt representation objects (ExceptionCommunication, ExceptionHardware, ExceptionDBError, InterruptGateway).

Details on each class can be found in the literature [101].

4.4 A NOVEL SOFTWARE CLASS: THE ELECTROMAGNETIC COMPILER

The HyperSurface API, presented in the preceding section, described the process of interacting with a HyperSurface from the aspect of a plain user. When a user requests the deployment of a virtual function on the HyperSurface, the corresponding configuration of the switch elements is extracted from the Configuration DB. A configuration extrapolation service handles cases not readily present in the DB. Then, proper communication commands set the state of the switch elements accordingly.

Thus, the compiler software that will be presented here can be described as the process of populating the Configuration DB with the required information/configuration entries. The term "compiler" is used to denote the general process of translating a high-level user input to low-level machine instructions.

4.4.1 A Qualitative View of the Compiling Process

The HyperSurface Compiler relies on multiple systems working together to produce the optimal configuration of a given metasurface

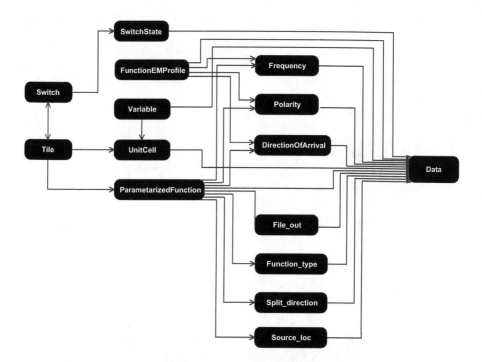

Figure 4.10 The HyperSurface API Class diagram.

device for a requested EM functionality. For example, the Compiler provides the answer to questions like: *"How must a HyperSurface be configured (e.g., what controller voltage settings must be used) so that it perfectly absorbs incoming radiation of 5 GHz at 45° incidence?"*

The systems employed by the Compiler include small- and large-scale full-wave EM simulations, simplified semi-analytical models, optimizers, anechoic chamber 'real-time' measurements and database storage and retrieval. The Compiler overview can be summarized in the following workflow steps:

1. Definition of the metasurface tile (unit-cell number and type, available configurations, etc.) and requested EM functionality (type, frequency, polarization, directions, etc.).

2. Preparatory EM simulations on a unit-cell (or super-cell) scale, i.e., without simulation of the entire metasurface.

3. Prediction of optimal configuration using semi-analytical models based on unit-cell-level simulation data (step #2) and appropriate fitness functions.

4. Optimization, iterating on steps #2 and #3.

5. Validation using full-scale (entire metasurface) EM simulations.

6. Automated anechoic chamber measurement(s).

7. Optimization, iterating on step #6, using appropriate fitness functions.

8. Store optimal configuration(s) in the database for future access.

Evidently, the full workflow presented above can only be performed on the "expert" side, i.e., by the designers of the HyperSurface. The "end-user" will mainly rely on the database but will also have (restricted) access to the full compilation workflow.

This section will mainly deal with step #3, i.e., the analytical models for predicting the optimal configuration through fitness function calculation. We will also present the required definitions and the underlying theory. These methods are effectively used in the optimization steps (#4 and #6), crucial to producing the best configuration. Overall the aim of this section is twofold: firstly to give formal definitions of "fitness" of a particular HyperSurface configuration for a particular (supported) functionality, and, secondly, to provide insight on how the HyperSurface must be configured to optimize the aforementioned fitness.

4.4.2 Metasurface Functions

Reconfigurable metasurfaces can be used for the manipulation of wavefronts of impinging waves. The classification of all possible manipulation is the set of available metasurface functions (or functionalities), each one parameterized. The more versatile the metasurface, the broader the supported functions spectrum. Broadly speaking, a metasurface can operate in a transmission and/or in a reflection mode or regime; in this deliverable, we will only focus on the latter, i.e., the reflection regime.

Two basic functionalities targeted by any metasurface are ABSORB and STEER. The former is the absorption of an impinging wave by the metasurface, while the latter is the reflection (steering) of the impinging wave at a desired direction, obviously different from the specular ("regular") reflection direction. In principle, these functions are applicable to arbitrary wavefronts (impinging and reflected), but for clarity's sake we will focus only on plane wavefronts. Figure 4.11 illustrates these three functionalities in a 2D side-view and for an impinging wave of TE polarization.

Other functionalities that can also be accomplished by the Hyper-Surface include:

1. REFLECT (referring to specular reflection, can be considered either as the complimentary of ABSORB or as an elementary STEER operation),

2. POLARIZE (change the polarization of the reflected wave, e.g., from linear to circular),

3. FOCUS (an impinging plane wavefront is transformed to a converging wavefront, focused at a given point),

4. COLLIMATE (the reciprocal of FOCUS, i.e., diverging spherical wavefront transformed to a plane wave),

5. BEAMSPLIT (separate the reflected wave in a number of beams at distinct directions, different from the specular reflection direction),

6. SCATTER (isotropically diffuse impinging radiation so that reflection to the specular direction is minimized, without absorption),

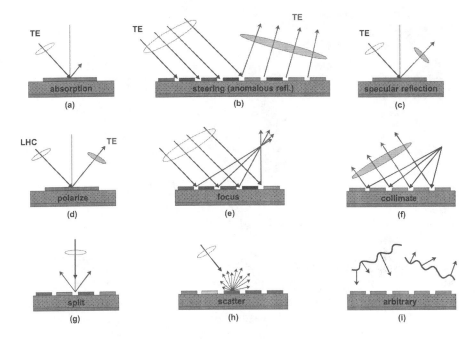

Figure 4.11 Wavefront manipulation types supported by metasurfaces, modeled as function types.

7. ARBITRARY (apply a frequency filter to one of the above functionalities, e.g., ABSORB only at a specified frequency; this functionality is implicitly applied by all the above functionalities).

Finally, it is needless to stress that all metasurface functions critically depend on the operation frequency and polarization, apart from the "angular" dependence outlined in Fig. 4.11.

4.4.3 Formal Definition of a Metasurface Configuration

In this subsection, we introduce a mathematical representation of the *configuration* of the metasurface tile. A configuration is a collection of specific switch states of the HyperSurface. As such, it can be seen as the output part of a Configuration DB entry. Figure 4.12 depicts a generic metasurface in top-view (left inset), where the brown rectangles represent meta-atoms, the green lines represent switch elements, and the pink background represents the dielectric substrate.

Let **s** be a single configuration, comprising the states of all switch elements on the HyperSurface in array form. Each s_{ij} element of array **s**

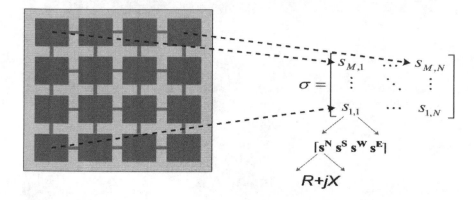

Figure 4.12 Formal definition of a Metasurface configuration.

corresponds to the active elements contained within meta-atom i,j. For this generic metasurface, each meta-atom may have one active element at each position "N", "S", "W", "E". Therefore, each element s_{ij} of **s**, as depicted in Fig. 4.12, is essentially a sub-array of size 1 x 4 as follows:

$$s_{ij} : \begin{bmatrix} \mathbf{s}^N & \mathbf{s}^S & \mathbf{s}^W & \mathbf{s}^E \end{bmatrix}.$$

Each element **s** (in bold) is a complex number representing the impedance of the corresponding physical active element. In the form of the impedance, **s** has the form:

$$\mathbf{s} : R + j \cdot X,$$

where $R \geq 0$ is the resistive and $X \in \mathbb{R}$ is the reactive part of the complex valued impedance controlling the unit-cell. R and X constitute the variables of configuration **s**. R and X have no other restriction: it can be unique for each s_{ij}/**s** element or identical over all, and each s_{ij}/**s** may have its own specific R, X values, or be fully depended on the R, X values of another element within **s**. The domain of R and X can freely be discrete or continuous (with/without bounds). Note that the complex impedance $R + j \bullet X$ can be related to a complex scattering parameter value, assuming decoupled ports and a given characteristic impedance, e.g., 50 Ω. Implemented ("real world") metasurface device configurations are usually designed and measured using the Touchstone file format, which contain S-parameters over a spectral range, for a given "real world" configuration.

The s_{ij}/**s** representation is intended to be used as a generic way of describing the positioning and active element configurations of a wide range of metasurfaces. As such, not every $s_{ij}/\mathbf{s}$ needs to be defined in each specific metasurface design case. For instance, the simple metasurface design shown in Fig. 4.12 can be described by using the $\mathbf{s}^N$ and $\mathbf{s}^W$ elements only, setting $\mathbf{s}^S$ and $\mathbf{s}^E$ to null ($\emptyset$). Moreover, the meta-atoms at the top and left sides of the HyperSurface have their $\mathbf{s}^N$ and $\mathbf{s}^W$ elements set to $\emptyset$ respectively.

In the following, we will assume that the HyperSurface has been properly defined by:

1. employing the proper $\mathbf{s}$ elements and setting $\emptyset$ where needed,

2. defining the independent R, X variables and their domains,

3. defining the derivation of any dependent R, X variables.

That said, we define Σ as the set of all possible configurations $\mathbf{s}$, i.e., $\mathbf{s} \in \Sigma$. For instance, a globally tunable binary metasurface has only two configurations: all elements "on" or all "off". A locally tunable on/off metasurface of 10×10 unit-cells will have 2^{100} possible configurations (including all permutation symmetries). Evidently, a locally tunable metasurface where the $\mathbf{s}$ value of each unit-cell can have continuous values will have an exponentially increasing number of possible configurations.

4.4.4 Definition of Fitness Function

Let a parameterized function of interest, f, as defined in the Hyper-Surface API. For instance, consider a function of type STEER, parameterized to reflect a direction of arrival $\overrightarrow{D_1}$ towards a direction $\overrightarrow{D_2}$:

$$f \colon \text{STEER } (\overrightarrow{D_1}, \overrightarrow{D_2}).$$

Moreover, consider a *Fitness* function that measures how good a configuration σ is for the parameterized function f. In other words, a fitness function maps function/configuration pairs $\{f, \sigma\}$ to a strictly ordered value set, e.g.:

$$\textit{Fitness}(f, \mathbf{s}) \; [\text{Value}_{low}, \text{Value}_{high}], \; (\text{e.g., } [0, 1] \text{ or } [-1, 1] \text{ or } [-30, 0] \text{ dB, etc.})$$

The value set can be freely defined as discrete, continuous, bounded or unbounded. The only restriction imposed is that the sorting of the values should strictly correspond to a monotonous fitness of configuration σ for parameterized function f.

With no loss of generality, we will make the convention that higher values correspond to better fitness and equal values correspond to equally good configurations.

Then, we define the optimal configuration σ_{best} as the one derived via the relation:

$$\sigma_{best} argmax_{\sigma \in \Sigma} \{Fitness\ (f,\ \sigma)\}.$$

Evidently, the purpose of the HyperSurface compiler is to populate the configuration DB with the entry $\langle f, \sigma_{best} \rangle$ for the particular metasurface tile. Thus, the configuration DB will eventually contain all pairs $\langle f, \sigma_{best} \rangle$ for each parameterized function of interest, making them readily available to end-users.

It is noted that the above optimization problem can be solved analytically only for simple parameterized functions, e.g., simple steering for a very specific input/output direction. Calculating the optimal configuration for any kind of complex parameterized function (e.g., beam splitting under bandwidth restrictions) necessarily requires a heuristic optimization approach in the general case. The heuristic optimization process can benefit from analytical insights, in order to speed up the optimization process and/or reach a more efficient outcome. For instance, in common STEER functions the Ohmic load parameter R should be zero, to minimize the absorption and maximize reflection. Moreover, for certain metasurface configurations, e.g., ABSORB from a specific direction of arrival and frequency, the real and imaginary parts of the impedance, R and X, are equal across all active elements.

Finally, as stated in the description of the HyperSurface API, extrapolation can be considered over the $\langle f, \sigma_{best} \rangle$ pairs contained in the configuration DB, in order to deduce configurations for parameterized functions that are not contained in the database.

4.4.5 Methods

In the next subsection, we define and calculate the fitness functions per metasurface function type (e.g., ABSORB, STEER), accompanied with common restriction objectives. We present two definitions of the fitness functions, calculated by different methods; these can be used separately, complementary or sequentially, during the metasurface configuration.

The first method relies on the **S-parameters** (reflection and transmission coefficients) extracted from a **Floquet-mode analysis** [63], based on full-wave EM simulation of a unit-cell or a supercell (a small group of unit-cells). Its merit lies in the significantly reduced computational effort in the simulations required, as only a fraction of the metasurface is actually modeled and simulated. This method can be applied to the fitness function calculation of some elementary functions (e.g., ABSORB, REFLECT, POLARIZE and STEER) but, most importantly, it serves as a basis for calculation of more elaborate wavefront shaping functions (e.g., FOCUS/COLLIMATE, SPLIT, SCATTER, etc.).

The second method relies on the **farfield radiation/scattering pattern** produced when the entire metasurface is illuminated by the incident radiation of interest. In principle, the illuminating wavefront can be arbitrary, but for simplicity's sake we will focus on a few special cases with practical applicability: plane waves, spherical waves and Gaussian beams. Nevertheless, the metrics defined in this deliverable can be used by the HyperSurface Compiler to optimize a metasurface even for arbitrary excitation, given enough resources (simulation and/or measurement time). It should be stressed that extracting the scattering pattern with full-wave EM simulations is very computationally intensive as the entire metasurface needs to be modeled. However, semi-analytical formulations can be derived that utilize Fourier optics together with the S-parameters calculated from unit-cell-level full-wave EM simulations. Now, in terms of the fitness function calculation, the scattering pattern method is more realistic and closer to the practical implementation, Moreover, these calculations are also applicable to an experimental testbed, i.e., when the Compiler has access to an automated measurement system (e.g., anechoic chamber).

As described in the Compiler workflow, we use both methods, S-parameters and scattering pattern: Firstly, we use the unit-cell-based method for preliminary calculations (where applicable), we then employ the semi-analytical scattering-pattern method for estimating the optimal metasurface configuration (if necessary), we subsequently employ the full-metasurface simulation for validating the configurations and, finally, we move on to measurement-based optimization and validation (also relying on the scattering-pattern method).

4.5 THEORETICAL FOUNDATIONS OF THE ELECTROMAGNETIC COMPILER

In this section, we will outline the framework required for expressing the metasurface functionalities in fitness metric form, mostly focusing on the targeted STEER and ABSORB functions, but also giving the fitness functions for other functions as well. We will start with the definition relying on the S-parameters that are extracted from a Floquet (or unit-cell) simulation of the metasurface, which assumes an infinite (or very large) number of unit-cells [63]. Subsequently, we devote a section to the fitness metric forms definition in terms of the farfield scattering/radiation-patterns, extracted when the full metasurface is simulated, semi-analytically modeled or experimentally measured. The scattering pattern method can in principle handle all possible functions, so it is the most general.

As described in the Compiler workflow, the Floquet analysis, requiring the EM simulation of only a small part of the metasurface, is meant to be used in conjunction with the semi-analytical models, which sometimes are trivial, e.g., as in the case of the ABSORB function. In this way, we can efficiently approximate the metasurface configuration required for a desired function. Subsequently, once the optimal unit- or super-cell configuration has been identified, a simulation of the full metasurface can be performed to calculate the far-field scattering/radiation-pattern and assess the deviations of the design performance when the infinite Floquet structure is truncated to produce the real structure.

The fitness metrics will be expressed in the normalized range $f = [0, 1]$ implying a [worst, best] fit, respectively. In case the optimizer, under use, works best in minimizing a fitness metric (instead of increasing it), and a logarithmic scale is preferable, then the metric can be expressed as $f' = 10\log_{10}(1 - f)$, where a value $f' < -30$ dB is considered acceptable.

4.5.1 Definitions

We adopt a standard spherical coordinate system where directions are specified by a pair of angles (θ, φ): the elevation angle ($\theta \in [0, \pi]$) origin is from the z-axis, normal to the metasurface, and the azimuthal angle ($\varphi \in [0, 2\pi]$) origin is the x-axis. The plane of incidence (PoI or incidence plane) is defined by two vectors: the normal to the metasurface (typically $\hat{z}$) and the direction of incidence. For instance, the plane of incidence in Fig. 4.13 is the xz-plane. Incident wave direction angles are given in the "inward" convection (i.e., by placing the tip of the direction vector at the point of incidence), while the reflected/scattered direction angles are in the "outward" convention (i.e., the end of the direction vector lies on the point of incidence). For example, the direction of the incident wave in Fig. 4.13(a) is $(\theta_i, \varphi_i) = (\sim 40^o, 180^o)$, while the reflected wave direction is $(\theta_r, \varphi_r) = (\sim 40^o, 0^o)$. This pair of angles, (θ, φ), is the information contained in the direction vectors $\overrightarrow{D_1}$ and $\overrightarrow{D_2}$ defined in other sections of the document, also referred to as DoA (direction of arrival).

Finally, note that we primarily consider plane waves propagating in air, i.e., transverse electromagnetic (TEM) waves. So, the polarization, as defined by their E-field vector, which is normal to the propagation direction, can have TE and/or TM components: TE (or "s") is the component normal to the plane of incidence and TM (or "p") is parallel to it. Given the complex-valued Cartesian components of the incident E-field, $\overrightarrow{E}_i = E_{xi}\hat{x} + E_{yi}\hat{y} + E_{zi}\hat{z}$, its TE- and TM-polarized components are calculated as follows:

1. Normalized vector of direction of incidence: $\hat{v}_i = \sin\theta \, \cos\varphi \, \hat{x} + \sin\theta \, \sin\varphi \, \hat{y} + \sin\theta \, \hat{y}$

2. Vector, normal to the plane of incidence: $\overrightarrow{v}_n = \hat{z} \times \hat{v}_i$

3. Unitary normal vector: $\hat{v}_n = \overrightarrow{v}_n / \|\overrightarrow{v}_n\|$ (where $\|\overrightarrow{v}_n\|$ is the Euclidean norm of a real-valued vector)

4. Unitary vector parallel to the PoI (and normal to $\hat{v}_i$): $\hat{v}_p = \hat{v}_n \times \hat{v}_i$

5. TE component of incident field: $\overrightarrow{E}_{i,TE} = \left(\hat{v}_n \cdot \overrightarrow{E}_i \right) \hat{v}_n$

6. TM component of incident filed: $\overrightarrow{E}_{i,TM} = \left(\hat{v}_p \cdot \overrightarrow{E}_i \right) \hat{v}_p = \overrightarrow{E}_i - \overrightarrow{E}_{i,TE}$

The fraction of TE and TM polarization components in the incident field can be calculated by the scalar values: $a_{TE} = \left\| \overrightarrow{E}_{i,TE} \right\|^2 / \left\| \overrightarrow{E}_i \right\|^2$ and $a_{TM} = 1 - a_{TE}$, where $\left\| \overrightarrow{E} \right\|$ is the Euclidean norm of a complex-valued vector.

4.5.2 Floquet (Unit-Cell) Analysis

In order to optimize our EM simulation workflow and minimize the computational effort, the ABSORB and STEER functions will be expressed in the context of Floquet (unit-cell) theory. This methodology consists of enclosing an elementary structure, referred to as unit-cell, within the so-called Floquet boundary conditions. These boundary conditions emulate an infinite periodicity of the unit-cell along the direction normal to the specified boundaries, implemented with

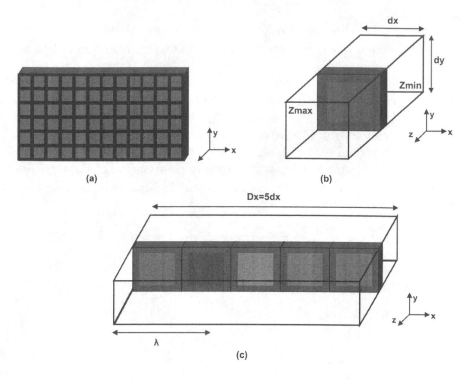

Figure 4.13 Metasurface unit cell and super cell illustrations.

imposed phase relations between each pair of opposing boundaries. This methodology requires a rather small computational domain and thus enhances simulation efficiency, as only a small fraction of the meta-surface needs to be modeled.

HyperSurfaces are planar, lying in the xy-plane (normal to z-axis), and consist of a rectangular grid of unit-cells arranged parallel to the x- and y-axes, Fig. 4.13(a). The dimensions of each unit-cell, Fig. 4.13(b), in the xy-plane are d_x and d_y, and the metasurface contains N_x by N_y unit-cells; in the example of Fig. 4.13(a), the metasurface contains a 10x10 grid of unit-cells, which is a relatively small number used for illustration purposes. The unit-cells usually have sub-wavelength dimensions, but, as will be discussed later on, this is not a prerequisite for applying the Floquet analysis. In fact, unit-cells can be defined in whichever manner, as long as they are rectangular and their periodic repetition builds up the metasurface; for example, in Fig. 4.13(b), the unit-cell is a square containing a metallic patch which is connected to a ground plane with a lumped element (not displayed).

To model these types of rectangular-grid planar metasurface, the unit-cell is isolated and Floquet (or unit-cell) conditions are applied to its four sides, normal to x- and y-axes, while its top and bottom sides (normal to z-axis) are considered as "ports" where scattering parameters (S-parameters, SP) can be calculated. A generalization of the reflection & transmission coefficient, the SP_{ij} parameter is a complex value (amplitude and phase) referring to the ratio of the scattered wave at the i-th port of a network over the incident wave at its j-th port, assuming all other ports are matched (i.e., terminated at loads with $Z_{in} = Z_0$, where Z_0 is the free-space wave impedance in the case of the metasurface studied). In our analysis, one or multiple ports of the Floquet-bounded unit-cell, i.e., of the implied infinite structure, are excited by an incident plane wave and the S-parameters are calculated at all the ports of the structure. So, given the unit-cell structure, the direction of the incident wave and its polarization, one can computationally solve for the EM behavior of the metasurface in a given frequency range.

The outputs of this EM simulation are the S-parameters in the ports of the unit-cell, i.e., at its top and bottom sides, and for specified "Floquet modes". Floquet modes can be thought of as the distinct ways in which the unit-cell can scatter EM radiation; in most practical applications, Floquet modes are plane waves propagating toward discrete directions. For instance, assuming that "$A = Zmax$" is the top port of the unit-cell, Fig. 4.13(b), and the impinging wave is TE polarized and propagates along the $(\theta_i, \varphi_i) = (30^o, 0^o)$ direction, the $SP_{(A,TE)(A,TE)}$ parameter refers to the plane wave reflected to the $(\theta_r, \varphi_r) = (30^o, 180^o)$ direction, as implied by specular reflection. Maximizing this parameter, e.g., $|SP_{(A,TE)(A,TE)}| \to 0$ dB, at a given frequency or frequency range, would mean that the metasurface perfectly reflects the wave coming from the given (oblique) direction. Inversely, minimizing this parameter, e.g., $|SP_{(A,TE)(A,TE)}| < -30$ dB, implies than reflection is practically zero, i.e., the unit-cell is "matched" to free-space propagation, which can mean that the incident wave is transmitted and/or absorbed by the metasurface, with no back-reflection.

Now, in case the unit-cell is not sub-wavelength, Fig. 4.13(c), i.e., when its x- and/or y-extent is larger than the operating wavelength [e.g., $D_x = 5d_x > \lambda$ in Fig. 4.13(c)], then the unit-cell can also support Floquet modes or "diffraction orders" with $(m, n) \neq 0$, where $(m, n) = (0, 0)$, the fundamental diffraction order, corresponds to specular reflection. These higher order Floquet modes correspond to plane waves reflected by the unit-cell in directions on either side of the specular, and can thus be used to perform steering (anomalous reflection). Electrically large unit-cells are referred to as "super-cells", and are typically blocks of non-identical unit-cells whose periodic repetition in both lateral directions forms the metasurface.

Assuming that the super-cell is larger than one wavelength in its x-dimension ($D_x > \lambda$) and the PoI is the xz-plane ($\varphi_i = 0^\circ$ or 180°), then the governing equation for the diffraction mode elevation angle (i.e., the angle inside the plane of incidence) is

$$\theta_{r,n} = \sin^{-1}\left(-n\lambda/D_x - \sin\theta_i\right),$$

stemming from momentum (wave vector) conservation. Equivalently, diffraction modes can arise also in the y-direction, when $D_y > \lambda$ and the PoI is the yz-plane ($\varphi_i = \pm 90^\circ$),

$$\theta_{r,m} = \sin^{-1}\left(-m\lambda/D_y - \sin\theta_i\right).$$

For instance, Floquet modes of $(n,m) = (\pm 1, 0)$ diffraction order w.r.t. {normal incidence ($\theta_i = 0$) and $D_x = 2\lambda$, $D_y < \lambda$} refer to plane waves propagating at angles $\theta_r = \sin^{-1}(\pm 0.5) = \pm 30^\circ$, i.e., directions $(\theta_r, \varphi_r) = (30^\circ, 0^\circ)$ and $(30^\circ, 180^\circ)$, respectively, in our spherical coordinate system. Note that when the argument of the inverse sine functions in the previous formulas is larger than 1 (in absolute value), the reflection angle becomes a complex number with real part equal to 90°. This means that the respective Floquet mode ceases to be a radiating (plane wave) mode and becomes evanescent.

Now, in the most general case where the incident wave direction is arbitrary (the only restriction that applies is $\theta \geq 90^\circ$, so that we have incidence from the top side of the metasurface) and the unit-cell size is D_x-by-D_y, then the reflected wave direction at a given diffraction order (n, m) is given by:

$$\theta_r^{(n,m)} = \sin^{-1}\left\{k_0^{-1}\sqrt{k_{x,r}^2 + k_{y,r}^2}\right\}, \qquad (4.1)$$

$$\varphi_r^{(n,m)} = \text{angle}\left\{k_{x,r} + jk_{y,r}\right\}, \qquad (4.2)$$

where $k_{x,r} = -k_0 \cos\varphi_i \sin\theta_i - 2\pi n/D_x$ and $k_{y,r} = -k_0 \sin\varphi_i \sin\theta_i - 2\pi m/D_y$. These are the components of the reflected wave vector that are parallel to the metasurface, and thus affected by it. It is evident that for appropriate values of D_x and D_y these equations can give rise to Floquet modes (i.e., reflected plane waves) at "anomalous" non-specular directions, e.g., $\theta_r \neq \theta_i$ and/or steering outside the PoI. The procedure outlined is the operation principle of the diffraction gratings, widely used in many applications of optics and microwaves.

It is of critical importance to stress that the spectrum of directions where reflected Floquet plane-wave modes can be directed is not continuous, but discrete, and the available values are a function of the wavelength (λ), the super-cell lateral dimensions (D_x and D_y) and the diffraction orders (n, m). One extra criterion is that the reflected angle must satisfy $\text{Re}\{\theta_r^{(n,m)}\} < 90^o$ in order for the (n, m)-th diffraction mode to be radiated and not evanescent.

Employing the Floquet mode simulation of a super-cell with the methodology described, for a given incident wave {direction, polarization, frequency}, we can calculate the S-parameters to higher order Floquet modes. For instance, in a super-cell with $D_x > \lambda$, the maximization of the parameter $SP_{(A,TE,10)(A,TE,00)} \to 0$ dB would mean that the super-cell "anomalously reflects" the impinging wave at the direction of the diffraction order $(n, m) = (+1, 0)$. For example, in the previous example, if $\theta_i = 0^o$ then $\theta_r = 30^o$ (and $\varphi_r = 0^o$).

In conclusion, employing the Floquet theory, unit-cells and super-cells can be used to optimize a metasurface for ABSORB and STEER functionalities, respectively, without having to EM simulate the entire metasurface. When the unit-cell dimensions are sufficiently small compared to the wavelength (e.g., $d_x, d_y \leq 0.1\lambda$) and the metasurface contains a relatively large number of them (e.g., $N_x, N_y \geq 50$), then the results of such a Floquet analysis will be practically identical to the ones acquired by a simulation of the full metasurface. The results will be reasonably accurate even for more relaxed values, e.g., $d_x, d_y \approx 0.2\lambda$ and/or $N_x, N_y \approx 25$, with observable deviations from the "perfect" case which cannot, in principle, be fully compensated.

4.5.3 ABSORB Functionality

Drawing from the presented Floquet theory, a metasurface can be designed so as to fully absorb an impinging wave of given polarization, direction of incidence and frequency, by maximizing the absorption in a single unit-cell. When that task is accomplished, it can be readily shown that a metasurface composed of a reasonably large number of these unit-cells will fully absorb the wave of the specified parameters {polarization, direction, frequency}.

The procedure is as follows: Initially, we assume that the unit-cell is sub-wavelength (i.e., its x- and y-dimensions are smaller than the operating wavelength) so that it will only produce specular reflection, and the wave is impinging from port "$Zmax$", e.g., from the top of the structure in, Fig. 4.13(b). Under these conditions, perfect absorption (PA) of the impinging wave is accomplished when both the reflection and transmission coefficients are zero. These coefficients can be directly related to the S-parameters of the unit-cell, defined for perfect absorption as:

$$|r| = |SP_{(Zmax,M)(Zmax,M)}| = 0,$$

and

$$|t| = |SP_{(Zmin,M)(Zmax,M)}| = 0.$$

In the previous notation, M represents a specific Floquet mode of the unit-cell defined by its:

1. direction of incidence, i.e., angles (θ, φ) in a spherical coordinate system,

2. linear polarization, i.e., TE or TM ("s" or "p"), if the wave is normal or parallel to PoI, respectively,

3. operating frequency (or wavelength, λ).

It can be inferred that the PA condition will strictly apply only for a single $\{\theta, \varphi, \lambda\}$ set, but absorption will also be significant in the proximity of that set, i.e., at a narrow frequency band or angular spectrum. Additionally, mixed polarization states, e.g., elliptical or arbitrary-linear, require a decomposition in the two orthogonal polarization components (TE and TM), and so lead to distinct reflection & transmission coefficients that cannot, in general, be both zeroed-out simultaneously.

Having stated the prerequisites, the fitness metric of the ABSORB functionality, for a given $\{\theta, \varphi, \lambda\}$ and polarization, can be expressed as:

$$f_{ABSORB} = 1 - \left(|r_{TE}|^2 + |t_{TE}|^2\right) - (|r_{TM}|^2 + |t_{TM}|^2),$$

which is in the range $[0, 1]$, denoting zero and full absorption, respectively. Note that this equation accounts for cross-polarization terms, i.e., when an incident wave containing one polarization only (e.g., TE) gives reflection/transmission to both polarizations (e.g., TE and TM).

The expressions of the fitness metric function can be simplified as follows: Firstly, if the unit-cell is terminated at its *"Zmin"* port [bottom of the unit-cell, Fig. 4.13(a)] by a fully reflecting surface, e.g., a PEC (perfect electric conductor) condition in simulations or a metallic backplane in practice, then the transmission coefficients can be safely assumed as zero, $|t| \to 0$, and excluded from the calculation of f_{ABSORB}. Secondly, if the unit-cell is symmetric under $\pi/2$ rotation around the z-axis, then cross-polarization terms are negligible, and can also be excluded from the calculations. In this case, the expression simplifies to

$$f_{ABSORB} = 1 - |r_{pi}|^2,$$

where $pi = \{TE, TM\}$, is the incident wave polarization.

4.5.4 REFLECT Functionality

Complimentary to ABSORB, the REFLECT functionality can be expressed as

$$f_{REFLECT} = |r_{TE}|^2 + |r_{TM}|^2 ,$$

in the range of $[0,1]$, denoting zero reflection (full absorption and/or transmission) and full reflection, respectively, for both the polarization components (TE and/or TM) present in the incident wave.

4.5.5 POLARIZE Functionality

The "polarize-by-reflection" functionality can be expressed by the fitness metric form

$$f_{POLARIZE} = |r_{pr}|^2 ,$$

where $pr = \{TE, TM\}$, is the desired polarization state in the reflected wave. In this simplistic notation, it is implied that the incident wave polarization is either a mix of both polarizations or only has the opposite polarization (e.g., $pi = TM$ and $pr = TE$). Moreover, the aforementioned function applies only to linear polarization states. When the desired polarization in the reflected wave is elliptical (e.g., left-hand circular), one also needs to apply a fitness metric to relative magnitude *and* phase of the transmission coefficients between the two polarizations (TE and TM), taking into account the incident wave polarization as well.

The generic polarization transformations are easily described in terms of the **Jones calculus**, i.e., using the matrix & vector formalism. Given the plane of incidence (PoI), the impinging and reflected plane waves are decomposed in TE ("s") and TM ("p") components, each one with its phase and magnitude; this is the Jones vector representation of a plane wave polarization. A metasurface can in principle treat the two polarizations differently (e.g., if its unit-cells are not invariable under 90° rotation around the normal vector), and this "treatment" is encapsulated in the Jones matrix of the metasurface. The overall polarization transformation operation is described as:

$$\overrightarrow{E}_{out} = [J]\,\overrightarrow{E}_{in} \rightarrow \begin{pmatrix} |E_{TE}|\,e^{j\varphi_{TE}} \\ |E_{TM}|\,e^{j\varphi_{TM}} \end{pmatrix}_{out} = \begin{bmatrix} J_{ss} & J_{sp} \\ J_{ps} & J_{pp} \end{bmatrix} \begin{pmatrix} |E_{TE}|\,e^{j\varphi_{TE}} \\ |E_{TM}|\,e^{j\varphi_{TM}} \end{pmatrix}_{in}.$$

According to the complex values of the Jones matrix $[J]$, the metasurface can act as a polarizer (linear or circular) or a retarder (quarter- or half-wave plate). Additionally, as planar metasurfaces usually have rectangular unit-cells arranged in the xy-plane, the Jones matrix elements are usually calculated (or measured/characterized) in terms of the $\hat{x}$- and $\hat{y}$-polarizations; consequently, one would need to decompose the impinging (in) and reflected (out) plane waves in these components, instead of the TE and TM; this is rather trivial in normal incidence or when the PoI coincides with a principal plane of the Cartesian coordinate system, but it can easily be extended for arbitrary oblique incidence. In any case, the fitness function metric for POLARIZE can be expressed as the Euclidean distance ("complex norm") of the actual outgoing Jones vector from the desired Jones vector:

$$f_{POLARIZE} = \left\| \overrightarrow{E}_{out} - \overrightarrow{E}_{desired} \right\|.$$

Finally, we should note that for polarizing plane waves, the whole metasurface must be configured homogeneously, i.e., all unit-cells must have the same configuration. Consequently, in order to optimize the metasurface, the calculation of the Jones matrix of a single unit-cell is required, which is easily done using the S-parameters.

4.5.6   STEER Functionality

As it was mentioned in the Floquet analysis, steering (also called anomalous reflection) can only be effectuated by super-cells, i.e., unit-cells larger than the operating wavelength in their lateral dimensions. Metasurface super-cells are composed of multiple sub-wavelength unit-cells, as in Fig. 4.13 which depicts a metasurface configured by a periodic repetition of a super-cell consisting of 5x1 unit-cells, for illustration purposes; note that in practical applications the super-cells are composed of a higher number of unit-cells.

According to diffraction grating (Floquet) theory presented, steering (anomalous reflection) of a plane wave coming from an incidence direction (θ_i, φ_i) by a super-cell is governed by Eqs. (4.1) and (4.2). Moreover, it was shown that the spectrum of directions where anomalous reflection can be directed is discrete, and the available directions are a function of:

1. the incidence direction (θ_i, φ_i),

2. the wavelength, $\lambda = 2\pi/k_0$,

3. the super-cell lateral dimensions, $D_x = N_x d_x$ and $D_y = N_y d_y$, and

4. the diffraction orders (n, m).

One extra criterion is that the reflected angle must satisfy $Re\left\{\theta_r^{(n,m)}\right\} < 90^o$ in order for the (n, m)-th diffraction mode to be a radiated one (not evanescent). Note that the polarization (TE or TM) does not affect the steering direction, only the complex value of the result reflection coefficient.

In theory, very large super-cells and/or very large diffraction orders can achieve steering towards any direction, but in practice the size of the super-cell is limited so deviations from the targeted steering directions will have to be tolerated. This means that the steering efficiency will be optimal at a wavelength close to the desired one, but not exactly on it; inversely, for a given wavelength, the incident plane wave will be reflected at an angle close to the desired one, but not exactly on it. In order to quantify these offsets, a tolerance in steering direction is defined $(\delta\theta, \delta\varphi)$.

The procedure for the optimization of STEER functionality is as follows:

1. Given parameters: wavelength λ, incidence direction (θ_i, φ_i), desired steering direction $(\theta_{st}, \varphi_{st})$, unit-cell dimensions (d_x, d_y).

2. Define tolerance in the steering direction $(\delta\theta, \delta\varphi)$, due to discrete nature of diffraction modes.

3. Iterate through the discrete set of values $\{n/N_x, m/N_y\}$, first by increasing N_x, $N_y = \{1,2,3,..\}$, and then $n, m = \{\pm 1, \pm 2, \dots\}$. Calculate the "candidate" direction (θ_r, φ_r) from Eqs. (4.1) and (4.2).

(a) **IF** $Re\{\theta_r\} < 90°$ **AND** $|\theta_r - \theta_{st}| < \delta\theta$ **AND** $|\varphi_r - \varphi_{st}| < \delta\varphi$, **THEN** the desired values of $\{N_x, N_y\}$, i.e., the super-cell size, have been found and the reflection coefficient corresponding to the "optimal" diffraction order mode (n_o, m_o) has to be maximized to optimize the steering.

(b) **ELSE**, continue iteration until a valid set of values is found. If the iteration leads to very high values for $\{n, N_x, m, N_y\}$, then steering for these parameters is not possible. One solution is to increase the tolerance $(\delta\theta, \delta\varphi)$.

4. Define the super-cell and apply the required reflection-phase profile across its constituent unit-cells. This profile is typically a linear gradient from 0 to $u2\pi$, where $u = \{m_o, n_o\}$, and is achieved by adjusting the reactance of each (non-identical) unit-cell in the super-cell.

5. Optimize the super-cell (e.g., tune the reactance of its constituent unit-cells) to maximize the reflection coefficient with EM simulations. These are performed using Floquet-bound super-cells and excitation from the top side of the structure, port $Zmax$, Fig. 4.13(b),(c). The polarization of the incident wave is also taken into account here.

Assuming the metasurface unit-cell dimensions and operating wavelength are defined in advance, the iteration in step #3 of the outlined procedure can be performed once, *a priori*, for a coarse grid of direction pairs (θ_i, φ_i) & $(\theta_{st}, \varphi_{st})$, and tabulated for future reference. A further simplification of this problem is when the unit-cell is square, $d_x = d_y$, and/or when the incidence plane is one of the principal Cartesian planes (xz- or yz-plane, i.e., $\varphi_i = \{0°, 90°, 180°, 270°\}$).

Elaborating on step #4 of the outlined procedure: Having identified the desired set of $\{n, N_x, m, N_y\}$ values and assuming that the super-cell is designed so that transmission is negated (e.g., by incorporating a metallic back-plane in the design), the super-cell is subjected to the Floquet simulation for the calculation of the S-parameters at a number of diffraction orders, typically $|n| \leq |n_o|$ and $|m| \leq |m_o|$. For the purposes of STEER functionality, the anomalous reflection coefficient at the incidence port of the super-cell, e.g., $Zmax$ (top side), is sought to be maximized:

$$|r_{pnm}| = |SP_{(pnm)(p00)}| \to 1,$$

where p is the polarization of the incident mode (TE or TM). Taking into account that the structure (super-cell) might also absorb a portion of the incoming wave as well as reflect it in other diffraction modes and polarizations, the most rigorous definition of a fitness metric for this functionality, bound in $[0, 1]$, is

$$f_{STEER} = 1 - |r_{\bar{p}n_om_o}|^2 - \sum_{p}^{TE,TM} \sum_{n \in M_x}^{n \neq n_o} \sum_{m \in M_y}^{m \neq m_o} |r_{pnm}|^2$$

where $\bar{p}$ is the cross-polarized mode of the same order as the desired one (e.g., $TM_{n_om_o}$ if the desired is $TE_{n_om_o}$) and M_x, M_y is the set of all diffraction orders considered in the Floquet simulation. In other words, if the super-cell does not absorb the impinging wave, then $f_{STEER} \rightarrow 1$ (optimal value) only when the reflection in all diffraction modes, except the desired one, is minimized.

Finally, two more notes should be made on the Floquet analysis. Firstly, that it is *reciprocal* w.r.t. the incident and reflected plane wave directions. For example, if a super-cell is optimized for the $\theta_i = 0^o \rightarrow (\theta_r, \varphi_r) = (30^o, 0^o)$ anomalous reflection case, it will identically function for the $(\theta_i, \varphi_i) = (30^o, 0^o) \rightarrow \theta_r = 0^o$ case. Secondly, that a super-cell optimized for anomalous reflection between the specular direction and a diffraction order ± 1, will also exhibit optimized *retro-reflection* to/from the direction corresponding to diffraction order ∓ 1. For instance, in the super-cell of the previous example, the wave incident from $(\theta_i, \varphi_i) = (30^o, 0^o)$ will be reflected back to $(\theta_r, \varphi_r) = (30^o, 0^o)$ with high efficiency.

4.5.7 SPLIT Functionality

The last metasurface function that can be modeled in terms of the S-parameters (Floquet mode analysis) is beam splitting, where an impinging plane wave is reflected and split in two or four plane waves traveling in directions symmetrically spaced from the specular reflection. Similar to the STEER function, SPLIT is also a product of diffraction grating theory: the metasurface is configured in periodically repeated supercells whose dimension is larger than the free-space wavelength; in this manner higher diffraction orders (modes) are supported. It can be shown that for a binary (two-level) reflection phase profile across the unit cells of the super-cell, e.g., half of them configured to $\Delta\varphi = 0$ and the other half to $\Delta\varphi = \pi$, the magnitude of the reflection coefficients corresponding to the first $(m, n = \pm 1)$ diffraction modes will be maximized while the reflection coefficient corresponding to the specular mode $(m, n = 0)$ will be minimized. In fitness metric form, this is expressed in terms of the reflection coefficient of the diffraction orders/modes as:

$$|r_{pnm}| = |SP_{(pnm)(p00)}| \rightarrow 0.707,$$

for splitting in two beams, which assumes that m (or n) is zero and $n = \pm 1$ (or $m = \pm 1$). For splitting in four beams, where $m = \pm 1$ and $n = \pm 1$, we require:

$$|r_{pnm}| = |SP_{(pnm)(p00)}| \rightarrow 0.5.$$

A less strict set of metrics, which also takes absorption into account, is the minimization of the specular reflection coefficient, $|r_{p00}| \rightarrow 0$, together with the equality of the other reflection coefficients, e.g., $|r_{p0(+1)}| = |r_{p0(-1)}|$ for splitting in two beams.

As previously stated, SPLIT is easiest to attain in binary reflective metasurface configurations, so that the simulation of only a relatively small super-cell is required. Moreover, it should be noted that the fitness calculation with S-parameters is accurate in normal incidence or when the plane of incidence coincides with a principal plane and the polarization is linear and parallel to one of the transverse axes of the metasurface plane ($\hat{x}$ or $\hat{y}$).

4.5.8 Far-Field Scattering/Radiation Pattern

Simulating a metasurface of a limited number of unit-cells, e.g., 25-by-25, illuminated by a plane-wave source (or a limited-width beam) of given direction of incidence can produce the radiation scattering pattern, which provides information on how the metasurface scatters the impinging wave in the far-field, e.g., when measured in a sphere of radius much larger than the scatterer dimensions normalized over the wavelength of operation.

4.5.9 Formal Definition

The radiation pattern can be expressed in terms of the radiation intensity $U(\theta, \varphi)$, which denotes the power per unit solid angle (in steradians) scattered towards the direction (θ, φ), in a spherical coordinate system using the "outward" pointing convention (end of direction vector positioned at point of incidence). Note that we employ the standard (ISO) spherical coordinates convention where $\theta = 0$ defines the $+\hat{z}$ axis, $(\theta, \varphi) = (90°, 0°)$ defines the $+\hat{x}$ axis and $(\theta, \varphi) = (90°, 90°)$ the $+\hat{y}$ axis. Equivalently, one can express the scattering pattern in terms of scattered power density $W_{sc}(\rho, \theta, \varphi) = \rho^{-2} U(\theta, \varphi)$ or scattered field amplitude $|E_{sc}| \propto \sqrt{W_{sc}}$ (in V/m), assuming a sphere of given radius (ρ). All these representations are in essence equivalent in terms of the direction(s) where scattered radiation is maximized or minimized or when estimating the lobe widths in the pattern.

4.5.10 Semi-Analytical Calculation

Using the principles of geometric and Fourier optics, assuming linear polarization in one of the principal metasurface axes ($\hat{x}$ or $\hat{y}$) and analytical defined source types (e.g., plane or spherical waves [point source]), we can express the E-field scattered from the metasurface with the following expression:

$$E_{sc}(\theta, \varphi) = \sum_{m=1}^{M} \sum_{n=1}^{N} A_{mn} e^{j a_{mn}} \cdot f_{mn}(\theta_{mn}, \varphi_{mn}) \cdot \Gamma_{mn} e^{j \phi_{mn}} \cdot f_{mn}(\theta, \varphi) \cdot e^{j k_0 \delta},$$

with $\delta = m d_x \sin\theta \, \cos\varphi \, - n d_y \sin\theta \, \sin\varphi$ and where:

1. θ, φ is a direction in the 3D scattering pattern, in our case the upper hemisphere above the metasurface,

2. d_x, d_y is the size of the unit-cells in the tangential directions,

3. M, N is the total number of unit-cells in the metasurface, along its two tangential directions, i.e., in the xy plane,

4. mn is the 2D numbering scheme used to identify a single unit-cell in the metasurface (e.g., row and column index),

5. $A_{mn} e^{j a_{mn}}$ is the illuminating field (amplitude and phase) impinging on the mn unit cell,

6. $\Gamma_{mn} e^{j \phi_{mn}}$ is the reflection coefficient (amplitude and phase) of the mn unit cell, stemming from its configuration for given polarization and frequency,

7. f_{mn} is the elementary scattering pattern of a given unit cell, typically assumed isotropic ($f_{mn} = 1$) or, for better accuracy when considering patch-like unit cells: $f_{mn}(\theta, \varphi) = \cos\theta$,

8. $\theta_{mn}, \varphi_{mn}$ is the direction of the source relative to the mn unit cell; we assume that only one "ray" from the source impinges on each unit cell.

Using this formula we can quickly approximate the scattering pattern from $E_{sc}(\theta, \varphi)$, assuming that we are given the impinging wave source (frequency, type, position, etc.) and the reflective metasurface configuration (number and size of unit-cells, each one's reflection coefficient). Simple source types include infinite plane waves, point sources (spherical waves) and Gaussian profiles, but in principle any wavefront can be treated, provided that its amplitude and phase on the metasurface are given.

This semi-analytical scattering pattern calculation can be further accelerated by noticing that the double summation can be reformulated into a 2D discrete Fourier transform (DFT), and making use of ultrafast FFT2 algorithms.

For some simple cases, e.g., that of an infinite plane wave ($A_{mn} = const.$) normally incident ($\theta_{mn} = 0$, $a_{mn} = const.$) on a reflective metasurface ($\Gamma_{mn} = 1$) of square unit cells ($d_x = d_y = d$), this expression can be greatly simplified:

$$E_{sc}(\theta, \varphi) \propto \cos\theta \sum_{m=1}^{M} \sum_{n=1}^{N} e^{j\phi_{mn}} \cdot e^{jk_0 d \sin\theta \, (m\cos\varphi - n\sin\varphi)}.$$

It can be inferred that the scattering pattern is effectively "shaped" by the reflection-phase profile applied across the metasurface unit-cells, ϕ_{mn}. In the previous example case, if we want to steer the radiation in the form of a plane wave towards a direction (θ_r, φ_r) the required phase profile, i.e., the reflection phase in each mn unit cell, is

$$\phi_{nm} = k_0 d \sin\theta_r \, [m\cos(\varphi_r + \pi) - n\sin(\varphi_r + \pi)].$$

Generalizing this concept, it can be shown that the reflection phase profile is the difference of the "requested" outgoing (reflected) phase front, ψ_{mn}, from the incoming (incident) phase front, a_{mn}, on each unit cell:

$$\phi_{mn} = \psi_{mn} - a_{mn}.$$

Please note that, in the previous calculation, the phase is assumed wrapped within $[0, 2\pi]$. That means that, ideally, the metasurface unit cells must be able to cover the entire reflection phase span, $[0, 2\pi]$.

To illustrate an example application, we will assume the COLLIMATE function, where a point source emits a spherical wave which impinges on the metasurface. The incoming (incident) phase profile on the metasurface, a_{mn}, is simply given by the geometrical distance of each mn unit cell from the point source, multiplied by the free-space wave number k_0.

$$a_{mn} = k_0 \sqrt{(x_s - x_{mn})^2 + (y_s - y_{mn})^2 + z_s^2}$$

where (x_s, y_s, z_s) is the location of the point source, and $(x_{mn}, y_{mn}, 0)$ is the center of each tile. Now, the requested metasurface function is to COLLIMATE this diverging beam, i.e., parallelize/steer all rays to the same direction (θ_r, φ_r), effectively forming a plane wave. As previously shown, the outgoing phase profile of such a plane wave is

$$\psi_{nm} = k_0 \sin\theta_r \left[md_x \cos(\varphi_r + \pi) - nd_y \sin(\varphi_r + \pi) \right],$$

and the required phase gradient to apply to the metasurface is

$$\phi_{mn} = mod\left(\psi_{mn} - a_{mn}, 2\pi \right).$$

This reflection phase profile corresponds to different reactance values (e.g., different capacitance) applied to each unit-cell of the metasurface via the controller switch fabric (the chips embedded in the metasurface). This correspondence is typically approximated by "looking up" the values (phase↔capacitance, for a given frequency and polarization) in the data pregenerated by unit-cell-level full-wave simulations, and then refining them.

Figure 4.14 presents such a test case of a COLLIMATE function. The point source (black dot) and the scattering pattern showing a clear plane wave lobe steered towards $(\theta_r, \varphi_r) = (40°, 0)$ can be seen on the left panel. The required metasurface phase profile for this particular collimation function can be seen on the right panel.

4.5.11 Polarization

Note that the polarization of the incident and scattered waves are interrelated, even though sometimes different conventions are used for each one. Incident polarization is customarily given in terms of the E-field components that are normal (TE or "s") and parallel (TM or "p") to the plane of incidence (PoI); thus, when the PoI coincides with one of the principal planes of the 3D Cartesian system, e.g., $x0z$ or $y0z$, then the TM/TE components are simple to translate to Cartesian components; in the case of oblique incidence, polarization is decomposed firstly in TE & TM and subsequently on the Cartesian axes, xyz, as these are the ones where a metasurface is typically characterized.

Full-3D scattering patterns are not restricted to the PoI or the principal planes and are always defined on spherical coordinates. Customarily, two scattering patterns are calculated, a co- and a cross-polarized

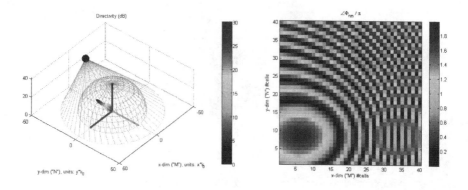

Figure 4.14 Example of COLLIMATE function: a point source (black dot on the left panel) emits spherical waves that impinge on the meta-surface configured with the reflection phases shown in the right panel; the produced scattering pattern is show in the left panel, exhibiting a clear lobe towards the $(\theta_r, \varphi_r) = (40^o, 0)$ direction.

with respect to the impinging wave polarization. Most frequently, we distinguish between the vertical $(\widehat{\theta})$ and horizontal $(\widehat{\varphi})$ polarization patterns, orthogonal between them, and defined with respect to the surface of flat earth ($\hat{z}$ pointing upwards). Yet another convention used, when the effective polarization of the structure is linear, is based on the two primary plane patterns, the E- and H-plane patterns, with respect to the plane containing the respective dominant field component.

Taking into account that radio waves in free space are TEM, the polarization of the impinging and scattered radiation in our spherical coordinate system will be expressed only in terms of the azimuthal $(\widehat{\varphi})$ and/or elevation $(\widehat{\theta})$ vectors. As the angles (φ, θ) depend on the spherical convention used, care must be taken when translating between coordinate systems, e.g., Cartesian to spherical, etc. Moreover, the desired 2D "slice" of the scattering pattern depends on the incident polarization and type of result representation (co/cross, vertical/horizontal, E/H plane). The aim of this discussion is to stress that polarization handling and representation are far from trivial, and to define that, unless otherwise stated, we will always consider the co-polarized scattering pattern representation, in 2D or 3D.

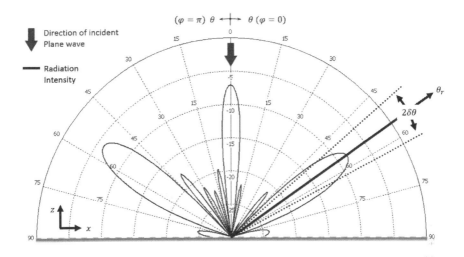

Figure 4.15 A metasurface scattering diagram.

4.5.12 Scattered Power in a Lobe (Solid Angle Cone)

The scattered power contained in a finite solid-angle lobe of the radiation pattern towards a given direction, e.g., (θ_r, φ_r) in the 2D pattern of Fig. 4.15, is given by the radiation intensity integral,

$$P_{scat}(\theta_r, \varphi_r) = \int_{\theta_r \pm \delta\theta} \int_{\varphi_r \pm \delta\varphi} U(\theta, \varphi) \sin\theta \, \mathrm{d}\varphi\mathrm{d}\theta,$$

where $\delta\varphi$ and $\delta\theta$ are the spans around the azimuthal and elevation angles where the integration is to be performed. These values can be thought of as the tolerances defined under the ABSORB case of the Floquet analysis. The total power scattered by the metasurface is given by the integration in the entire upper hemisphere, i.e., $\theta = [0, \pi/2]$ and $\varphi = [0, 2\pi]$.

$$P_{scat,T} = \int_{\theta=0}^{\pi/2} \int_{\varphi=0}^{2\pi} U(\theta, \varphi) \sin\theta \, \mathrm{d}\varphi\mathrm{d}\theta.$$

In what follows, we will define the fitness functions for the ABSORB and STEER functionalities, assuming one has access to the full 3D scattering pattern of the metasurface, for an impinging plane wave of specified {direction, polarization, frequency}. Fitness for the other available metasurface functionalities will also be outlined.

4.5.13 Fitness Functions per Functionality

We proceed to defined fitness functions per functionality, which can be used in a modular fashion within the compiler's optimization process.

4.5.13.1 ABSORB Functionality

Assuming the power of the incident plane wave is known, P_{inc}, the fitness metric for perfect absorption is extracted by the scattering pattern as:

$$f_{ABSORB} = 1 - \frac{P_{scat,T}}{P_{inc}},$$

and takes values in the range [0,1] (worst, best) when all the power is scattered in the upper hemisphere or when none of it is scattered (i.e., all is absorbed), respectively. Note that in these calculations, the polarization of the incident wave is not required, as we assume that the absorption is not polarization-selective, i.e., we seek to absorb both polarizations as much as possible.

In a practical experimental setup, where one seeks to optimize the fitness function, the following procedure should be followed: Firstly, all the unit-cells of the metasurface are set at the same configuration (e.g., the same $Z = R + jX$ values) so that only specular reflection is expected. Subsequently, the emitting and receiving antennas, at the appropriate polarizations, are placed in the same incidence plane and pointed at equal elevation angles, $\theta_i = \theta_r$ (specular reflection). The Z value that leads to perfect absorption will be the one for which the receiving antenna power is minimized; note that rotating the receiving antenna to get the full radiation pattern is not required. To fine-tune the Z value, searching in the vicinity of the value identified in the previous step, one can acquire the full radiation/scattering pattern to account for the power that is not absorbed, but scattered by the metasurface (to directions other than that of specular reflection). Finally, cross-polarization and wide-band measurements would ensure that there is no undesired "cross-talk".

4.5.13.2 STEER Functionality

Assuming the power of the incident plane wave is known, P_{inc}, and that we have defined a tolerance for the desired lobe solid angle, $(\delta\theta, \delta\varphi)$, the fitness metric for steering in the direction (θ_r, φ_r) is given by

$$f_{STEER} = \frac{P_{scat}(\theta_r, \varphi_r)}{P_{inc}}.$$

As in the absorb functionality, the fitness metric lies in the range [0,1], corresponding to a worst and best fitness, respectively. In the example case of Fig. 4.15, the values for the steering fitness function optimization would be $(\theta_r, \varphi_r) \approx (55^o, 0^o)$ and $2\delta\theta \approx 15^o$, and the estimated fitness value (assuming no absorption) would be $f_{STEER} \approx 0.25$; evidently, this structure exhibits poor steering efficiency in the desired direction and requires optimization.

Note that the formula presented above makes no mention to the polarization of the steered wave. As mentioned in the subsection pertaining to Polarization, scattering patterns can refer to the total field ("norm" of both orthogonal polarizations) or to only one polarization component (e.g., vertical or horizontal scattering pattern). In principle, when it comes to STEER, the outward direction is the same for both polarizations according to Floquet (diffraction grating) theory, and only the magnitude changes. So, once the fitness metric is evaluated for the total power, the polarization information of the steered wave can be extracted and a "polarization efficiency" factor can be applied to the previously calculated fitness metric.

In a practical experimental setup, where one seeks to optimize the fitness function, the following procedure should be followed: Firstly, one needs to program the $Z = jX$ values of the unit-cells so that an approximately correct phase gradient is imprinted on the metasurface, e.g., as in Fig. 4.15 for steering in the xz-plane. This procedure involves grouping the unit-cells in super-cells, and assigning appropriate capacitance values to each of them (e.g., by look-up in computationally extracted or measured curves for the phase of the unit-cell reflection coefficient as a function of capacitance, for near-zero resistance). Having prepared the metasurface configuration and positioned & aligned the transmitting and receiving antennas to the desired directions, the optimal $Z=jX$ values will be found by seeking to maximize the received power. Acquiring the full-radiation pattern would be used to make sure that the scattered power is also taken into account, i.e., that side-lobes at undesired directions are minimized. Cross-polarization and wide-band measurements would ensure that there is no undesired "cross-talk".

An example case of steering from $(\theta_i, \varphi_i) = (30, -45)$ to $(\theta_r, \varphi_r) = (80, 90)$ is depicted in Fig. 4.16. The metasurface is made of a 40 by 40 array of square unit cells (12 mm), the frequency is 10 GHz and the total polarization is considered.

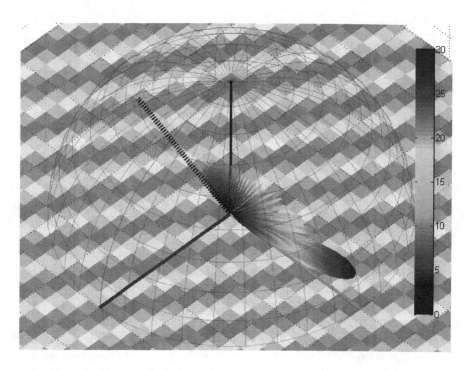

Figure 4.16 A 3D wave steering case.

4.5.13.3 REFLECT Functionality

The fitness for this metasurface function is identical to STEER, with the note that the steering direction is now given explicitly by the law of reflection rather than the diffraction grating formula.

4.5.13.4 SPLIT Functionality

Extending the REFLECT and STEER functions, in SPLIT we are looking for a maximization of the scattered power in multiple directions, (θ_n, φ_n), where $n = 1, 2, \ldots N_{lobe}$, i.e., the number of possible lobes. Typically N_{lobe} is two or four, when splitting the beam in one principal plane or both, respectively.

$$f_{SPLIT} = \frac{1}{P_{inc}} \sum_{n=1}^{N_{lobe}} P_{scat}(\theta_n, \varphi_n).$$

Note that the splitting directions (θ_n, φ_n) are determined by: the super-cell size (in both transverse directions), the operating wavelength and the incidence direction, as described in diffraction grating (Floquet mode) theory. Moreover, to optimize the SPLIT efficiency, i.e., maximize the fitness metric, "symmetric binary" configuration of the metasurface is prescribed: the supercell must have an even number of unit cells, where half of them are configured to reflection phase φ_0 and the other half to phase $\varphi_0 + \pi$.

Simply put, in order to split in two beams the metasurface must be configured in "stripes", whereas to split in four beams, a "checkers" configuration pattern is required. The transversal size and orientation of these stripes or checkers, determines the splitting directions. SPLIT example scattering patterns from a normally impinging plane wave are depicted in Fig. 4.17.

4.5.13.5 POLARIZE Functionality

POLARIZE requires knowledge of both complex valued (phase and amplitude) orthogonal field components, on the desired lobe/direction. Simulation software and/or measurement setups will typically produce such 3D (or 2D) scattering patterns, one for each polarization; each direction on each pattern will be attributed an amplitude and a phase value; correlating these and comparing them with the desired outgoing polarization will produce the fitness metric. For instance, given the co- and cross-polarized scattering patterns, $P_{co}(\theta, \varphi)$ and $P_X(\theta, \varphi)$, the desired polarization $\overrightarrow{E}_{out}$ (in terms of co- and cross-components), and

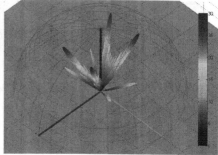

Figure 4.17 Wave splitting example in two (left) and four (right) directions.

the direction/lobe of interest (θ_o, φ_o), the fitness metric is:

$$f_{POLARIZE} = \left\| \vec{E}_{out} - \left(\begin{array}{c} P_{co}(\theta_o, \varphi_o) \\ P_X(\theta_o, \varphi_o) \end{array} \right) \right\|,$$

where we assumed that both Jones vectors are normalized (to unity norm), and that the complex values of scattered power P_{co} and P_X have been produced by appropriate integration in the cone $(2 \cdot \delta\theta, 2 \cdot \delta\varphi)$ given the angular spans/tolerances.

4.5.13.6   FOCUS and COLLIMATE Functionalities

We will first present COLLIMATE, whose aim is to "parallelize" all reflected rays coming from a diverging source towards a specific direction in 3D space (θ_c, φ_c). Similar to STEER and REFLECT, the fitness metric is simply

$$f_{STEER} = \frac{P_{scat}(\theta_c, \varphi_c)}{P_{inc}},$$

where we assume that the proper polarization pattern is considered.

FOCUS directs all rays of a collimated beam bundle towards a specific point in 3D space. It is reciprocal to COLLIMATE, meaning that the same metasurface configurations that optimize the COLLIMATE fitness between the two "conjugate" shapes (the point source and the beam direction) also optimize the FOCUS fitness provide that we switch input and output. Note that the scattering pattern is a far-field quantity that cannot produce meaningful quantities at the near-field (where the focal point is assumed to lie); in this regard, the reciprocity "transformation" is very useful.

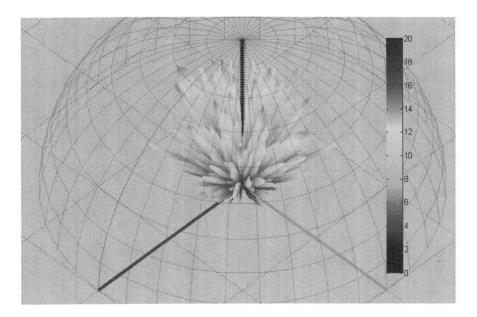

Figure 4.18 A scattering function performed by a HyperSurface.

4.5.13.7 SCATTER Functionality

The aim of this metasurface functionality is to "diffuse" the scattered radiation in all directions, e.g., as in Fig. 4.18. In a sense, this has the same end-effect with absorption, meaning that no radiation is reflected back towards the transmitter or in any direction in particular. This application is useful in RCS (radar cross-section) reduction, typically compared against the normally incident back-reflection from a perfect reflector, e.g., a metallic sheet. In order to optimize this functionality, the scattering pattern must be approximately isotropic in the upper hemisphere, i.e., $D_0 \to 2$ (max directivity of the pattern, assuming reflective [PEC-backed] metasurface); in practice, realistic scatterers will have a directivity much higher than 2 (e.g., 30-40) so the fitness metric can be defined as

$$f_{SCATTER} = \frac{2}{D_0} = \frac{P_{scat,T}}{\max{(U)}\, 2\pi},$$

where $\max\{U\}$ is the maximum radiation intensity in the upper hemisphere; we repeat that, in practice, $f_{SCATTER} > 0.1$ is considered optimal scattering.

As intuition implies, the optimal reflective metasurface configuration for SCATTER is when the reflection phase is randomly distributed along the unit cells. The following figure presents the scattering pattern of such a random configuration from normal incidence ($f_{SCATTER} = 0.05$).

4.5.13.8 ARBITRARY Functionality

Relying on the "blind" optimization algorithms employed by the Compiler, the end-user can define an arbitrary scattering pattern (for a given polarization and frequency) and several concurrent optimization goals, e.g., lobes & nulls in particular directions, directivity and sidelobe levels, etc. The optimizer will attempt to satisfy each goal with the prescribed "weights" and come up with the best configuration.

For instance, given the scattering patterns for vertical and horizontal polarizations: $P_{H,V}^{(t)}(\theta, \varphi)$ ("t" for target, desired) and $P_{H,V}^{(a)}(\theta, \varphi)$ ("a" for actual, e.g., measured) a generalized fitness metric definition is:

$$f_{ARBITRARY} = \sqrt{f_H^2 + f_V^2},$$

where the fitness for the horizontal and vertical scattering pattern matching are given by the weighted overlap integrals:

$$f = \frac{1}{1 + \langle e \rangle}, \ where \ \langle e \rangle = \frac{\iint_\Omega |P^{(t)} - P^{(a)}| |P^{(t)}| d\Omega}{\iint_\Omega |P^{(t)}| d\Omega}.$$

In this definition, $\langle e \rangle$ is the "weighted error", ranging from zero (best fit) to infinity (worst fit). This error is then normalized in the [0,1] range with the $(1 + \langle e \rangle)^{-1}$ operation, so that zero/unit is the worst/best fit. Note that the values of the scattering pattern functions $P(\theta, \varphi)$ are in principle complex and that the integration is performed on the solid angle hemisphere $d\Omega = \sin\theta d\theta d\varphi$, with $\theta = [0, 90]$ and $\varphi = [0, 360]$ (and "wrapped").

4.5.14 The Configuration Optimization Process

The solution to the optimization problem leading to the optimal $\langle f, \sigma_{best} \rangle$ pairs in the Configuration DB is derived by combining an optimization procedure with a fitness evaluation process, which can be either an experimental measurements setup or an electromagnetic simulator. The evaluation process can be seen as a function (in the generic sense) that measures the performance of a configuration **s** and outputs its fitness. The general workflow is depicted in Fig. 4.19.

The output of the optimizer drives the workflow by producing new candidate configurations σ' (i.e., the R/X values of the σ' array)

Evaluated fitness of σ'

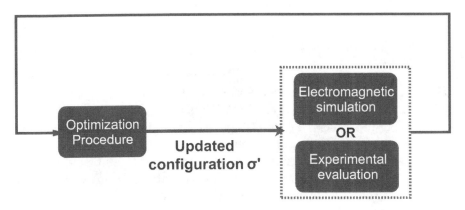

Figure 4.19 Function configuration optimization workflow.

towards a non-monotonic solution search, where the successively generated candidate σ' maps to varying fitness values, not all of them improving, but which over time provide a highly efficient trajectory towards the best configuration. The process continues until some termination criterion is satisfied. The latter is usually defined by a limit expressing the user's preference regarding:

1. The fitness level that can be deemed as satisfactory.

2. The amount of time to be devoted to the search.

When the configuration fitness evaluation is based on experimental measurements, the workflow is straightforward and realistic. First, a setup of emitting and receiving antennas (at the directions of interest) is implemented. Second, it is ensured that the measurement outcome is captured electronically at the device running the optimization service. Third, the optimizer begins to produce candidate configurations, which are deployed on the HyperSurface via its API and networking capabilities, described in Section 4.3. Finally, the corresponding fitness measurement is derived and is fed to the optimizer, repeating the cycle until a termination criterion is met.

When the configuration fitness evaluation is based on software-based electromagnetic simulations, the optimization cycle is identical. Nonetheless, the simulations incur runtime considerations, accentuating the need for efficient optimization processes and parallel computation capabilites. Moreover, analysis-assisted optimization and fitness functions are employed, as described in Section 4.5, dealing with the Floquet (unit-cell) analysis.

4.6 SOFTWARE ASPECTS OF THE ELECTROMAGNETIC COMPILER

This section contains the description of the various compiler-related software modules, in the form of UML diagrams.

It is noted that the described modules complement the ones described in Section 4.3 (API), but refer to a different user audience. Specifically, Section 4.3 referred to a more generic audience of software/system developers who seek to incorporate metasurfaces in their designs, (without requiring knowledge of the underlying physics). This section refers to the specialist that creates the necessary information, i.e., the tile function to tile configuration mapping, making HyperSurfaces accessible to an audience of non-specialists.

Therefore, in contrast with Section 4.3, the term "user" in the ensuing text refers to the specialist. The general developer audience is explicitly denoted as "general-user" where required. The contents of this section are as follows:

First, we describe the general use cases of the compilation-related software. These can be classified into Configuration DB maintenance and Configuration calculation tasks, which are studied in separate subsections. Then, we provide the Unified Modeling Language (UML) description of key-processes.

4.6.1 General Use Cases

Figure 4.20 presents the use cases covered by the compilation software. These include the tasks pertaining to the calculation of the optimal configurations per tile function (parameterized), as described in Section 4.5. Additionally, they include the tasks referring to the Configuration DB maintenance pertaining to the specialist-user. These exemplary include registering a tile by production serial number, and populating the DB with entries describing its structure (e.g., switch element types, locations and allowed states). Naturally, a general user should have this type of information available (and preferably in read-only mode). He can then extract the information referring to his owned tile production number and proceed to deploy configurations on it.

We describe the computational tasks first, and then the DB maintenance tasks.

Computational Task: Execute Optimization

This task refers to the derivation of the optimal configuration that best matches a required parameterized HyperSurface function (e.g.,

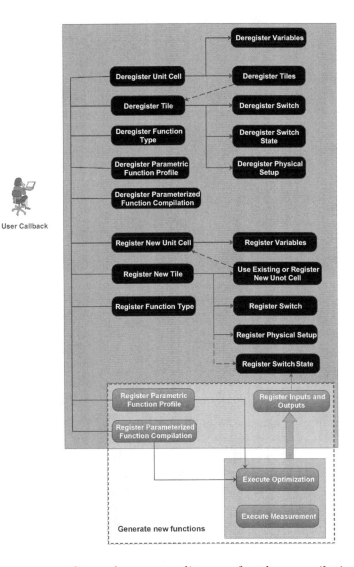

Figure 4.20 General use case diagram for the compilation-related software.

STEER from a given direction to another). Once triggered by the user, the optimization-driven process described in Fig. 4.19 is initiated, assuming that the proper information about the computational or experimental testbed resources have been provided. This task makes no differentiation between simulation-based and experiment-based evaluation. The user is expected to tune the software according to his intended approach.

Computational Task: Execute Measurement

The optimization task comprises a continuous cycle of measurements and corresponding configuration revisions until the optimal one has been found. Nonetheless, single measurements are also allowed, beyond the context of the optimization. The user may use this capability to freely check the behavior of the tile in some conditions of interest. More formally, the user may deduce a full profile of the electromagnetic behavior of a tile, which contains the behavior of the tile when it receives inputs (incoming waves) that are not intended by the presently deployed parameterized function. This information can be important for realistic applications.

DB Maintenance Task: Register/Deregister Unit Cell

This task refers to the process of initialing the unit cell, defined as the primary building block of the Tile. For each unit cell the number, type and position of the free variables are being stored. These variables are universal, representing a specific set of parameters that each switch element of a tile can be associated with. This offers an essential data structure that can be shared between different tiles.

DB Maintenance Task: Register/Deregister Tile

This task refers to the process of populating (or updating) the database with the necessary information for describing the structure of a tile. Each tile is uniquely defined by the type and number of unit cells composing its structure. All switch elements are being stored and associated with a free variable of the unit cell, defining the position and type of the element. The allowed states of the switches may also be stored and used as a constraint for the optimization process. This data configuration does not only provide a standardized way of representing and reusing optimization outcomes, but also the allowed input variable ranges for a future optimization task.

Notice that, as in the case of the API (Section 4.3), tile units are uniquely identified, e.g., on a serial number basis. This means that even tiles of the same general type (unit cell, geometry, metasurface pattern, switch elements) are represented with multiple entries, without sharing information. This approach is followed in order to facilitate the "capturing" (identification) of manufacturing artifacts that locally impair the device, e.g., by limiting the tunability range of a given switch on a specific tile. Moreover, it allows for storing fault detection information (e.g., switch hardware failure) on a per-tile basis.

Deregistering refers to the process of removing all information pertaining to a tile from the database (e.g., due to end of support or associated errors).

DB Maintenance Task: Register/Deregister Function Type

This tasks refers to the addition (or update) of a new supported function type (e.g., STEER, ABSORB, etc.). Here, supported function type means that the corresponding fitness function has been added to the compiler source code. Users can query the DB configuration for a list of supported function types.

Deregistering a function type results to the removal of all relevant data, including any existing optimal configurations.

DB Maintenance Task: Register/Deregister Parametric Function Profile

This task refers to the proper storage (or update) of the measurements outcomes of a single experiment or simulation, in the context of the EM profiling of a given metasurface function. This task specifies how this outcome should be handled in order to be stored in a standardized, re-usable manner.

The registration process also updates the stub tables of the DB (like polarizations, frequencies, DoAs, etc.) based on the inputs/outputs of the optimization process (sub-task "Register Inputs and Outputs").

The deregistration process removes the data for a given function profile.

DB Maintenance Task: Register/Deregister Parameterized Function Compilation

This task refers to the proper storage (or update) of an optimal configuration to the DB. The optimization process may return the results in a format that best fits the experimentation and simulation processes. This task specifies how this outcome should be handled in order to be stored in a standardized, re-usable manner.

The registration process also updates the stub tables of the DB (like polarizations, frequencies, DoAs [directions of arrival]) based on the inputs/outputs of the optimization process (sub-task "Register Inputs and Outputs").

The de-registration process removes the data for a given configuration compilation.

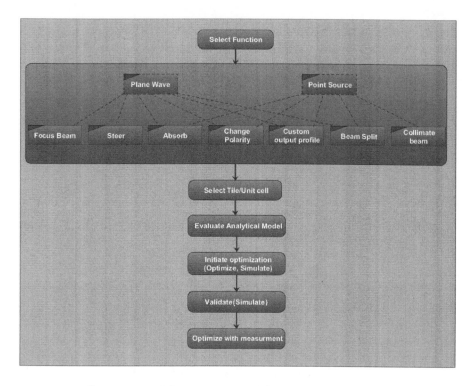

Figure 4.21 General workflow diagram for the optimization process.

The general procedure followed by this task is illustrated in the workflow diagram of Fig. 4.21.

A proper function can be selected between several options, each defined with either a plane wave or a point source as input. For each of the available options the optimization process can be performed either at a single unit cell or completely, for a specific tile. In particular:

1. All optimization requests for functions that allow a point source input execute at the unit cell level. In general, a point source is expected to individually illuminate each unit cell with a unique triplet for the magnitude, phase and polarization of the arriving EM field (Fig. 4.22). Therefore depending on the location of each cell in the grid, its optimization goals change accordingly. For instance a Tile with a grid of 40×40 unit cells is expected to designate a total of 1600 separate optimization goals.

2. For Plane Wave (PW) input the functions that require a full Tile approach are the following:

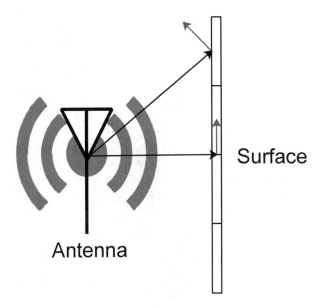

Figure 4.22 Semantic of point source illumination.

    (a)   i. A. Focus Beam
             B. Beam Split
             C. Custom profile

3. For Plane Wave (PW) input the functions that can operate with a single unit cell approach are the following:

    (a)   i. A. Steering
             B. Absorption
             C. Change Polarity

For the three functions belonging to the latter case, each optimization result is consistent with all tiles utilizing the same unit cell while the rest are always unique to the specific tile selected.

After the Tile or unit cell selection step is successful, an analytic evaluation based on the approximate formula detailed in Section 4.5 is ensued (Eqs. 4.1 and 4.2). The result returns the optimization goals for all unit cell/s of the selected Tile in the form of desired generalized scattering parameters for each polarization of the impending wave. The scattering parameters should be next translated to the proper R,C values for the switches belonging to all unit cells that have assigned optimization tasks. This constitutes the most challenging and time consuming part of the process and requires the combination of state-of-the-art optimization algorithms and the execution of several EM simulations.

After the completion of the latter step a full-tile simulation run is performed to validate the newly recommended R, C values. The process concludes with an experimental evaluation, if available. During this step a complementary optimization procedure can take place around the best-fit values of the former software optimization process. This allows a better convergence to a more realistic physical scenario, taking into account each minor physical detail of the tile and avoiding the constraints and assumptions of the numerical algorithms.

4.6.2   Validating the Compilation Outcomes with Measurements

A measurement task involves the precise control of the two primary units: a Network Analyzer and a Positioner apparatus, for a complete evaluation of the HyperSurface far field pattern. The general setup follows a standard bi-static measurement, as illustrated in Fig. 4.23. The Network Analyzer constitutes the main measurement device, responsible for transmitting and receiving the electromagnetic signal from and to both antennas inside the anechoic chamber. Altering the position of the HyperSurface is achieved through direct control of the Positioner. The motion of the tile is achieved over two degrees of freedom (rotational motions), namely the turn table and the mast angle as illustrated in Fig. 4.23.

The combination of the permissible values for these two types of motion can generate the complete rotational space required for acquiring the far field pattern. In other words, any relative position of the receiver and the transmitter with regard to the HyperSurface can be achieved.

Standard instrumentation connectivity protocols can be employed for the automated control of both the Network Analyzer and the Positioner units, such as the well-known GPIB protocol (General Purpose Interface Bus – IEEE-488), which is supported by the available experimental equipment. Employ a virtual star topology, a central GPIB server can receive scripted commands and delegate their execution to the required instrument. A standard connection is established from the HyperSurface API to the HyperSurface, and automation commands are send to rotate the testbed, following the specific syntax supported by the Network Analyzer and the Positioner units. Notice that any additional code can constitute a specific add-on to the compiler soft-

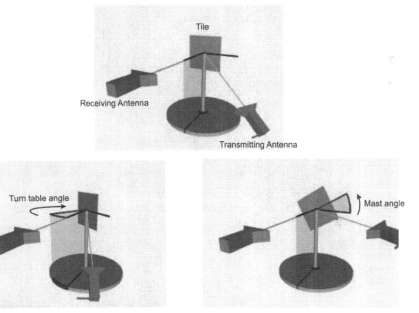

Figure 4.23 A setup for bi-static EM measurements to derive the scattering pattern of a HyperSurface configuration.

ware described in Section 4.6. However, the measurement process is not related to the HyperSurface API described in Section 4.1 and no code dependencies are added to it.

4.7 CONCLUSION

This chapter presented the HyperSurface software model, comprising an Application Programming Interface, a data model, and the electromagnetic Compiler. The latter is a novel class of software that seeks to translate high-level user directives (required EM behavior of the HyperSurface) to low-level actuation directives (corresponding states of the HyperSurface active elements).

This presented compiler sought to serve as a platform that can serve any kind of complex user directive that may be required in any application setting. To this end, the compilation process was formulated as a generic optimization problem. General categories of EM behavior were formulated as a fitness functions, which can subsequently be combined and optimized under custom restrictions.

An automated workflow was established to implement the compilation process, taking into account scalability (computational parallelism), portability between both simulated and experimental EM testbeds, as well as interdisciplinary software development facilitation considerations. Key-aspects of the HyperSurface software were described in detail, ranging from high-level use case diagrams to detailed class diagrams.

ACKNOWLEDGMENTS

This work was funded by the European Union's Horizon 2020 research and innovation program, Future and Emerging Topics (FET Open) under grant agreement No 736876.

CHAPTER 5

Design of the HyperSurface Networking Aspects

Dimitrios Kouzapas

Department of Computer Science, University of Cyprus, Nicosia, Cyprus

Taqua Khairy

Department of Computer Science, University of Cyprus, Nicosia, Cyprus

Nouman Ashraf

Department of Computer Science, University of Cyprus, Nicosia, Cyprus

Ileana Papailiou

Department of Computer Science, University of Cyprus, Nicosia, Cyprus

Anna Philippou

Department of Computer Science, University of Cyprus, Nicosia, Cyprus

Andreas Pitsillides

Department of Computer Science, University of Cyprus, Nicosia, Cyprus

Konstantinos Michail

SignalGeneriX, Limassol, Cyprus

Anastasis Kounoudes

SignalGeneriX, Limassol, Cyprus

CONTENTS

5.1 Design Requirements of the HyperSurface Controller
Network 143

5.2   HyperSurface Networking Components: The
HyperSurface Network Controller 144

     5.2.1   HyperSurface Controller Communication 144

5.3   The HyperSurface Controller Network Topology 145

     5.3.1   HyperSurface Network Controller Addressing 146

     5.3.2   HyperSurface Network Controller Channel
Mapping 146

5.4   HyperSurface Controller Network Communication
Protocols 148

     5.4.1   Routing and Reporting Protocol 148

     5.4.2   Fault-Adaptive Routing 149

     5.4.3   Workload Characterization 152

5.5   Evaluation of the Controller Network Design and
Performance via Simulations 153

     5.5.1   Custom-Built Simulations 153

     5.5.2   HyperSurface Controller Network Simulator 154

         5.5.2.1   The HyperSurface Controller Network
Simulation 154

     5.5.3   Formal Evaluation of the HSF-CN 156

     5.5.4   HyperSurface Emulator 159

5.6   The Controller-Gateway Communication Perspective 169

     5.6.1   Gateway Functionality 169

         5.6.1.1   Software/Firmware Design and
Development 177

         5.6.1.2   Tile Gateway Communication
Interface Firmware 179

         5.6.1.3   Error/Fault Detection 184

         5.6.1.4   Bluetooth Mesh Firmware 186

5.7   The HyperSurface within Control Loops 188

     5.7.1   System Model 189

     5.7.2   The Considered Model 190

     5.7.3   Control Algorithm 193

     5.7.4   Estimation Algorithm 193

     5.7.5   Performance Evaluation 195

5.8   Summary 196

HYPERSURFACES (HSFs) comprise structurally reconfigurable metasurfaces whose electromagnetic properties can be changed via a software interface, using an embedded miniaturized network of controllers, enabling novel capabilities in wireless communications, including 5G applications. The resource constraints associated with the hardware testbed of this breakthrough technology, necessitate an interconnected architecture of a Network of Controllers (HSF-CN) that is distinct from, yet reminiscent to, conventional architectures (e.g., Network-on-Chip (NoC) architectures).

To meet the requirements of the HSF hardware, the network is constructed via an irregular topology where network controllers are interconnected in a Manhattan-like geometry, with communication between controllers conducted using handshaking protocols [85, 127]. Routing within the network is operated by an XY-YX algorithm [34].

This chapter presents the HSF-CN selected topology, and the adopted communication protocols, including fault management protocols. The design is evaluated using complementary techniques. The chapter also introduces the design and implementation specifics of the HyperSurface Gateway (GW). The GW design and implementation includes the selected hardware and software technologies, as well as the protocols for sequencing routing packets toward the HSF-CN and fault management protocols.

5.1 DESIGN REQUIREMENTS OF THE HYPERSURFACE CONTROLLER NETWORK

The HSF configuration layer constitutes a set of interconnected controllers that are responsible to configure the meta-material atoms. The controller interconnection forms the HyperSurface Controller Network (HSF-CN) that must be characterized by scalability and robustness [85].

Scalability is dictated by the large number of meta-atoms that will be accommodated in practice: small meta-atom size is required for correct meta-material operation; and the need to avoid EM interference. In particular, the size of the meta-atom should be comparable to the incident wave wavelength λ (in the order of $\lambda/2$), while the metasurface thickness should be much smaller than the wavelength (in the order of $\lambda/10$). Scalability implies a simple controller design while the cost of each tile must be kept at a minimum. The potentially large surfaces to be covered, such as building walls, also necessitate designs of low power consumption.

Robustness is also required for the HSF-CN, i.e., the HSF-CN must be able to tolerate faulty controllers, especially in large scale HSFs that accommodate thousands of controllers. The CN architecture must offer reliable data delivery even in the presence of faults, foreseen due

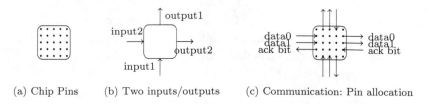

(a) Chip Pins   (b) Two inputs/outputs   (c) Communication: Pin allocation

Figure 5.1 Controller Node Communication Channels and Pin Allocation [85].

to imminent component failure(s), external influences such as accidental/intentional damage, and loss of connectivity.

5.2 HYPERSURFACE NETWORKING COMPONENTS: THE HYPER-SURFACE NETWORK CONTROLLER

The design of the HSF-CN is structured upon the design of the HSF-CN Controller. The HSF-CN controller is able to receive and handle software directives in the form of multiple network packets that contain configuration values. Moreover, the controller is able to use the configuration values and configure the meta-atoms under its responsibility. To accommodate the functionality of configuration packet handling in a robust manner, the HSF-CN controller implements three basic operations: configuration packet processing/consuming; configuration packet routing; and acknowledge the receipt of a configuration packet.

5.2.1 HyperSurface Controller Communication

Each controller needs 22 pins for the purposes of intra-tile communication, operations, and for meta-material configuration. The communication channel allocation of the controller is given in Fig. 5.1. The left diagram of the figure shows the available number of pins on the controller chip. The middle diagram shows the channel endpoint allocation for the controller. Each controller allocates two input channel endpoints and two output channel endpoints. The right diagram also shows the pin allocation for the four channel endpoints on the controller. At the hardware level [127, 128], each channel endpoint requires three signals for implementing asynchronous communication (Section 5.6.1.2). Figure 5.1 also shows the spatial distribution of the channel endpoints; on a clockwise direction there are two input endpoints followed by the two output endpoints.

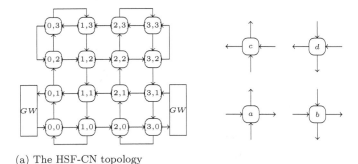

(a) The HSF-CN topology

Figure 5.2 Left: Manhattan Topology with bidirectional channel at the edges; Right: Controller Orientations.

5.3 THE HYPERSURFACE CONTROLLER NETWORK TOPOLOGY

Based on the controller design and channel allocation, the proposed topology for the HSF-CN is the Manhattan-like network topology with bidirectional channels at the edges as presented in Fig. 5.2 (left). To spatially achieve the connectivity (input/output endpoint connection) of a Manhattan topology the controller is rotated, with all four 90-degree rotations used to form the topology. Figure 5.2 shows the four spatial directions of the controller. The controller orientations are referred to with letters a, b, c, and d as shown in Fig. 5.2 (right).

The Manhattan topology is a grid structure network topology [116]. Due to the limited number of channel endpoints and the requirement for a grid structure layout of the controllers, the proposed topology has unidirectional communication both row-wise and column-wise, with alternating communication direction flow both row-wise and column-wise. What distinguishes the proposed Manhattan topology, however, from the one considered in the literature is the wraparounds: instead of connecting the opposite ends of a row/column, wraparounds connect neighbouring nodes on rows/columns where no connectivity is present. Our design choice of connecting neighbouring periphery controllers and not the ends of each row and each column is due to the hardware implementation: crossing the interconnection wires would require to add extra layers on the Printed Circuit Board (PCB) board that embeds the metasurface. Furthermore, the edge controllers would require components, e.g., transistors, with more signal drive to send signals over longer wires.

The adopted wraparound design, allows for connecting the *Hyper-Surface Gateway* by replacing a wraparound with a gateway device. The gateway device is responsible to connect the user and the HSF-CN, by processing and issuing software defined directives. As shown in Fig. 5.2 (left), for the purposes of the current HSF-CN design two gateways are connected to the Manhattan-like grid; one Gateway is to replace the wraparound between controllers $(0, 0)$ and $(0, 1)$. The output channel endpoint of the Gateway will be connected with the input channel endpoint of controller $(0, 0)$ and the input channel endpoint of the Gateway will be connected to the output channel endpoint of controller $(0, 1)$. This will allow for the Gateway controller to send packets directly to controller $(0, 0)$ and receive report packets from controller $(0, 1)$. A second gateway, that acts as a gateway for receiving acknowledgments, replaces the bottom right wraparound of the HSF-CN.

For the requirement of tiling, the topology is also chosen to contain an even number of controllers at the side of the grid. If an odd number of controllers is chosen, then it would not be possible to perform tiling and maintain the alternating sequence of the rows and columns of the multi-tile Manhattan topology in the larger formed topology.

5.3.1   HyperSurface Network Controller Addressing

Each controller in the HSF controller network needs to be uniquely identifiable via an addressing scheme. The addressing scheme used for the HSF controller network follows the Cartesian coordinates within the Manhattan-like grid as shown in the left diagram on Fig. 5.2 (left). The bottom left controller has $(0, 0)$ as its xy-coordinates. The x-axis increases from left to right and the y-axis increases from bottom to top.

5.3.2   HyperSurface Network Controller Channel Mapping

Output directions at the controller level are used to uniformly present the protocols within the Manhattan topology. This is done by mapping directions "up", "down", "left", and "right" with the output endpoints of each controller. The mapping is done based on the orientation, i.e., type, of the controller within the topology. The type of each controller is determined by its Cartesian address. Specifically, due to the alternate direction characteristic of the Manhattan topology, the type of the controller is determined by the x-coordinate, and respectively y-coordinate, modulo 2. Figure 5.3 shows a locally based routing scheme for a controller of type "a".

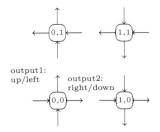

Figure 5.3 Local Routing Directions.

Type	Direction			
	Up	Down	Left	Right
"a"	output 1	output 2	output 1	output 2
"b"	output 2	output 2	output 1	output 1
"c"	output 2	output 1	output 1	output 2
"d"	output 2	output 1	output 2	output 1

Table 5.1 Mapping of Controller Outputs to Directions.

Specifically, the type "a" controller in the figure maps the routing directions "up", "down", "left", and "right" to its two output channel endpoints. Direction "up" denotes sending to a higher y-coordinate, and direction "down" denotes sending to a lower y-coordinate. Similarly, direction "right" denotes sending to a higher x-coordinate and direction "left" denotes sending to a lower x-coordinate. In the case of a type "a" controller direction "up" is mapped to output endpoint 1 and direction "right" is mapped to output endpoint 2. This mapping allows for a direct routing according to the specification of the "up" and "right" directions. Moreover, direction "left" is mapped to output endpoint 1 (just as direction "up") and direction "down" is mapped to output endpoint 2 (just as direction "right"). This is because routing in these cases cannot happen directly, rather it happens indirectly; type "a" controller indirectly routes a packet towards the "left" direction by first directing the packet to the "up" direction. Similarly, a type "a" controller indirectly routes a packet towards the "down" direction by routing the packet first towards the "right" direction. Table 5.1 shows the direct and indirect local level routing directions for all types of controllers.

5.4 HYPERSURFACE CONTROLLER NETWORK COMMUNICA-
TION PROTOCOLS

In the following, the HSF-CN communication protocols will be presented. More specifically, the HSF-CN supports protocols for: controller initialization; configuration packet routing; and reporting packet routing.

5.4.1 Routing and Reporting Protocol

Two main requirements for the HSF design are the packet routing protocol and the controller reporting protocol. The former is a protocol for routing data/configuration packets from a GW controller to a controller, whereas the latter is a protocol for routing report packets, e.g., acknowledgment packets or measurements information, from a controller to a GW. This section discusses the packet routing protocol and the controller reporting protocol.

The routing protocol adopted for implementation is a variant of the XY routing protocol [34]. The XY protocol in a monotonic topology [118] requires as parameters the target coordinates. Routing takes place as follows: the packet is routed on the x-axis until it reaches the x-column, then it is routed on the y-axis until it reaches the target. The proposed XY protocol, however, is adapted for the specific requirements of the Manhattan edge wraparound topology.

The XY routing procedure takes as a parameter the packet to be routed, and the coordinates of the target controller. In the Manhattan edge wraparound topology, the protocol, whenever possible, follows the same XY route as with the monotonic topology. However, due to the row-wise and column-wise alternate directions of the Manhattan edge wraparound topology, this is not always possible. In this case, the packet is routed indirectly to the target following the mapping of the directions "up", "down", "left", and "right". Figure 5.4 shows the paths that are used for routing packets towards controllers. Different colors in Fig. 5.4 represent different paths.

The XY protocol is chosen to be implemented because of its simplicity and deadlock freedom in conventional mesh topologies. The XY protocol allows for variant designs that can be implemented in future implementation iterations and offers flexibility and robustness. For example, an extension of the XY protocol, the XYi protocol, allows for multiple routing path options at the controller GW level. Each path is determined by an extra input that sets the routing column. A GW can compute, based on several factors, the best XYi routing path for a

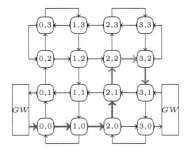

Figure 5.4 Routing Paths.

packet and set it through the routing column parameter. Factors that might be taken into account by the GW for choosing a routing path are: avoid faulty controllers, if any; reduce congestion, since different routing paths are expected to balance the network load, etc. [129]. The XY routing protocol is presented in terms of pseudocode in Table 5.2.

The XY protocol can be used to send packets from a GW to a controller or from a controller to a controller. To facilitate the reporting functionality, the XY protocol is also used to send acknowledgment packets from a controller to the acknowledgment Gateway. Acknowledgment packets are interpreted differently than configuration packets but they are routed using the same XY routing protocol. Each controller, after receiving a configuration packet, if requested it creates an acknowledgment packet. The acknowledgment packets' destinations are set to the maximum x value and the minimum y value to guarrantee that the acknowledgment packet will reach the GW at the bottom right corner of the HSF-CN, i.e., the acknowledgment Gateway. Using this kind of addressing for acknowledgment packets results in a scalable design of the HSF-CN that can accommodate grids of any size.

5.4.2 Fault-Adaptive Routing

Faults may appear in the controller network due to different reasons as for example component failure and external physical damages to the surface [154]. In any case, the routing protocol must have the ability to deliver packets to healthy nodes in the presence of faults. The uniqueness of the controller network topology and the constraints of the nodes dictate the development of new fault adaptive protocols for the considered network [85, 139]. Inspired by the Networks on Chips (NOCs) literature we develop two types of fault-adaptive routing protocols. Both protocols utilize the aforementioned XY routing mechanism as a

```
1   procedure Routing: packet
2
3       a, b: packet.x, packet.y
4       x, y: intratile controller coordinates
5
6       if (x == a and y == b)
7           terminate
8       end if
9
10      if (x/2 < a/2)
11          send packet right
12      end if
13
14      if (x ≥ a and y < b)
15          send packet up
16      end if
17
18      if (x > a and y ≥ to b)
19          send packet left
20      end if
21
22      if (x == a and y > b)
23          send packet down
24      end if
25  end procedure
```

Table 5.2   Routing Protocol.

baseline. The one relies on an ad-hoc strategy that blocks turns based on the locations of the current node, the faulty node, and the destination node, and it is referred to as the reliable delivery algorithm. The other utilizes victimization and turn prevention policies to route around faulty nodes as discussed below.

The reliable delivery algorithm is developed based on the principle that each source node can choose between two alternative, *disjoint paths*, dubbed $Path_A$ and $Path_B$, to route a packet towards a target node in the presence of faulty components. $Path_A$ initiates at one of the two output links (i.e., edges or channels) at the source node, and ends at one of the two input links of the destination node. Equivalently, $Path_B$ initiates at the alternative output channel of the source node, and terminates at the alternative input channel of the target node. Note that, in a similar manner, the oblivious XY routing algorithm along with its symmetric oblivious YX algorithm also comprises disjoint paths onto the proposed topology. An additional property of any of these route-traversing path pairs is that the first path spans the topology in a clockwise mode, while the second path spans it in a counter-clockwise mode.

Victimization-based routing relies on containing faulty nodes in convex rectangular shapes surrounded by rings of healthy nodes that facilitate routing around faults. This method combines victimization with turn prevention policies to ensure a provably fault-tolerant deadlock-free routing in the considered topology.

For the purposes of fault-tolerance in the current design of the HSF-CN a fault-tolerant protocol has been developed that follows a more naive approach. It calculates the set of nodes that are reachable with the XY algorithm, the set of nodes that are not reachable with the XY algorithm but are reachable with the Bellman-Ford or Dijkstra algorithm and the set of unreachable nodes (either because they are faulty or because there is no available path to reach them). Then when those sets are calculated the appropriate bypass packets are sent in order to bypass the faulty nodes. Its effectiveness is demonstrated via simulations conducted on a custom developed simulator using the Any-Logic Platform. The aim of the faulty tolerance (naive) algorithm is to bypass the faulty links in order to send configuration packets successfully through the topology. The algorithm iterates over the set of nodes in order to distinguish them in the three following sets: reachedXY (nodes that can be reached using the XY algorithm), reachedD (nodes that can be reached with the Dijkstra or Bellman-Ford algorithm) and unreached (nodes that cannot be reached). For each node it checks if both input links are faulty, then it updates the unreached nodes set. Then it calculates the path from the sendGW to the node (target) (pathToTarget) and the path from the node (target) to the ackGW (pathToEnd). If it is reachable with the XY algorithm (meaning that no faults were identified from the sendGW to the target node) then it updates the reachedXY set and if it is reachable with other algorithms it updates the reachedD set (meaning that faults were identified from the sendGW to the target node). In the following if there are no faults from the target node to the ackGW it means that ackGW can receive ack for the specific node (node.ack = true). Then it updates the list of packets with the packets and the respective bypass packets and returns the list. The algorithm terminates if the first edge is faulty or when the list of packets has been calculated.

5.4.3   Workload Characterization

To approach the design of the internal network of controllers properly, it is necessary to understand the workload that the internal HSF network will have to serve. In other words, we need to characterize the expected traffic between the HSF's controllers in order to make informed design decisions regarding the HSF network.

In this section we summarize a traffic characterization methodology of programmable metasurfaces, as overview in Fig. 5.5 (the reader is referred to [137] for details). Given the breadth of the problem, we first focus on a particular yet relevant functionality: anomalous reflection for beam steering. To this end, we develop a methodology to obtain the state of each unit cell as a function of several design parameters (e.g., unit cell size) and application parameters (e.g., steering angle). We apply the methodology to study the variations in the unit cell's states as functions of changes in the required reflection angle. Assuming that every state change implies sending a message, these variations provide spatiotemporal information on the metasurface traffic. By mapping angular changes to physical movements in real-world scenarios, our methodology can be applied to real use cases such as tracking a mobile user in a 5G network.

First an HSF of size $m \times n$ is assumed to track an object that can follow multiple mobility models. Without loss of generality, we consider horizontal movement parallel to the surface, projectile movement and intermittent random movement. As shown in Fig. 8.17, from the mobility models and the position of the HSF tile the incidence angles and the reflection angles of the incident and reflected waves, respectively, are measured. Then, through the metasurface coding method explained

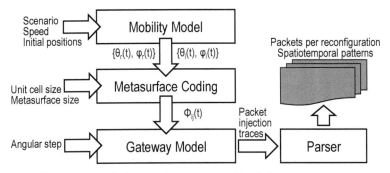

Figure 5.5 Summary of the evaluation methodology.

in [137], the phase gradient of the surface, from which the phases of all unit cells are calculated, is found. The phase gradient information combined with design parameters, such as the angular step, are utilized to obtain a matrix of unit cell states for a particular pair of incidence and reflection angles. Comparing consecutive matrices through a simple *diff* function determines the unit cells to which a reconfiguration request must be sent.

For each type of movement, we collect traces that describe the time instant at which packets are generated as well as their intended destinations. The traces are obtained and classified as a function of the following: (a) the number of states of a unit cell: the number of phase values a cell can obtain, and (b) the angular step: the smallest change in angle that requires a change in the HSF state. Finally, we parse the traces to obtain the percentage of state changing cells and the spatio-temporal distribution of traffic over the surface.

5.5 EVALUATION OF THE CONTROLLER NETWORK DESIGN AND PERFORMANCE VIA SIMULATIONS

The HSF-CN development process (design, implementation, integration) has deployed several methodologies that complement each other and provide with extensive results for the evaluation of the HSF-CN design and performance.

5.5.1 Custom-Built Simulations

At the initial design phase, simulations were developed as fast algorithm prototyping to check basic properties, e.g., packet delivery and deadlock-freedom, of alternative design choices (selected network topologies and protocols). The use of these simulations focused mostly on developing alternative deadlock-free designs and scenarios, rather than extracting useful measurements. After developing and checking a correct and deadlock-free design using these simulators, the design can be studied further using the other evaluation approaches. The topologies and protocols studied using this method can include more sophisticated routing protocols, which integrate fault identification/management techniques for future HSF-CN implementations.

In particular a custom-made Java simulator simulator was used as part of an iterative process to rapidly evaluate alternative routing and dissemination designs of the proposed pilot HSF-CN implementations. The simulator is also used for conducting rapid prototyping research that can be applied on future HSF-CN designs. Several versions of the simulator have been developed that implement different network configurations and routing algorithms. The custom-made simulator has also been used to evaluate more sophisticated routing protocols, fault identification, and fault tolerance techniques.

The basic requirement of the custom simulator is the delivery of packets from an entry controller to any controller within the grid. The packet contains data and the address of the recipient. Beyond correctness and deadlock freedom, the simulator also gathers statistical data gathered: the number of available routing paths towards the recipient, and their length measure in hops. This analysis leads to conclusions about shortest, average, and longest paths and their statistical distribution.

5.5.2   HyperSurface Controller Network Simulator

After the initial HSF-CN design was chosen, a network simulator was developed using the AnyLogic simulation tool [74], as a standard network simulative approach. This methodology allows the detailed simulation of basic networking aspects and it is used for evaluating a wide range of simulation metrics, which are used to validate the design against the project's requirements, such as throughput, delay, etc. The algorithms and techniques are normally evaluated using extensive simulation runs on various networking scenarios, but they are not exhaustive (i.e., they do not check against all possible application scenarios).

5.5.2.1   The HyperSurface Controller Network Simulation

AnyLogic is a Java-based multi-paradigm software development environment, which supports the following common modeling methods: i)

Discrete event; ii) System dynamics; iii) Agent-Based; iv) A combination of the above.

AnyLogic offers extra flexibility to the modeller with visual modeling languages like stock & flow diagrams, state charts, process flow charts and action charts. Simple models can be created using only drag and drop from the AnyLogic palette. More complex problems/behavior simulation, is done using the Java programming language. An AnyLogic model is mapped and translated in Java. Furthermore, AnyLogic offers a ready-to-use clock-less communication option as implemented in the HSF-CN. All the above features contributed towards selecting AnyLogic for the purpose of the HSF-CN simulation analysis.

The Agent-Based modeling method was used to create the HSF-CN network simulator. The simulator implements four different Agents: i) the Main Agent is the Top-level Agent, where the environment with the other agents is created; ii) the Controller Agent represents the network controller and is responsible for the routing/configuration process; iii) the Packet Header Agent contains all the information needed by the controllers for routing; and iv) the Packet Agent that represents the data bits of the configuration packet. AnyLogic supports synchronous, asynchronous, and mixed communication modeling. Asynchronous modeling is closer to real applications, so the conditional events, that give rise to asynchronous routing, are the main core of the HSF-CN simulator.

In Fig. 5.6, all the initialization parameters are illustrated. The first initialization process is based on the parameter size, which is considered as the size of the topology. The simulator creates the nodes and links, then connects the nodes with the links and builds the Manhattan edge wraparound grid. The second initialization process is based on the parameters traffic, ackx, acky, GWx, GWy. The input GW (GWx and GWy) determines where the packet will start and the acknowledgment Gateway (ackx and acky) the location of a successful termination of the packet. The traffic is an integer variable and it triggers the conditional event of a specific case we want to test on the simulator. These events are responsible for the traffic of nodes on the simulator, the pattern of packet target locations and the fault types. For example, we can decrease traffic within the grid or increase traffic by sending a packet whenever a controller is ready to receive. Finally, these events are responsible for the distribution of faulty controllers. The parameter FProbability is the probability that each individual controller can become faulty and we can simulate static faults during the simulation or dynamic faults, by allowing the probabilistic failure of controllers every time a new packet is created. The simulator supports both asynchronous and synchronous packet creation.

The simulator provides a graphical representation of the packet routing within the network. Traffic congestion, routing algorithm correctness, and packet paths can be easily inspected on the graphical interface rather than in text files. In addition, the advantage to pause or set the simulation speed and the runtime access to all the classes and variables significantly helps the detection and analysis of any routing problems. It can also generate and update graphs at run time.

5.5.3 Formal Evaluation of the HSF-CN

Formal verification is the process of employing formal methods to check whether a system meets its intended specifications (behavior). In the case of the developed HSF-CN, the behavior concerns the correct behavior of the developed algorithms for initialization, routing, and reporting. In contrast to testing and simulative methodologies, formal method techniques perform an exhaustive search of the state space of the model, thereby providing full guarantees of correctness.

Model checking is a popular approach towards formal verification [84]. It is fully automated and it is supported by model checking toolkits that have been developed and applied to various industrial settings, such as hardware verification and safety-critical systems. In model checking, a system is represented as a semantical structure or a model, and a property is encoded as a logic formula. The model-checking problem is to check whether the model satisfies the formula, thereby establishing the satisfaction of the property by the system. The key limitation of model checking is the state-space explosion problem: the state space of the system grows exponentially in the number of variables encoding the system.

In the context of our analysis, we employed the model checker tool UPPAAL [29] to exhaustively check the state space of particular HSF-CN designs. Such exhaustive checks of the executions led to the identification of subtle cases where bugs and/or deadlocks were present. The model checker also supports Statistical Model Checking techniques to approximate the distributions of metrics, e.g., the distribution of the time needed to cover the entire HSF, and alleviate the state-space explosion problem in the analysis of demanding scenarios (e.g., larger topologies).

The evaluation of the HSF-CN begins with the development of an UPPAAL model of the HSF. The model is a set of timed automata that describe the behavior of the controller and the GW. For per-

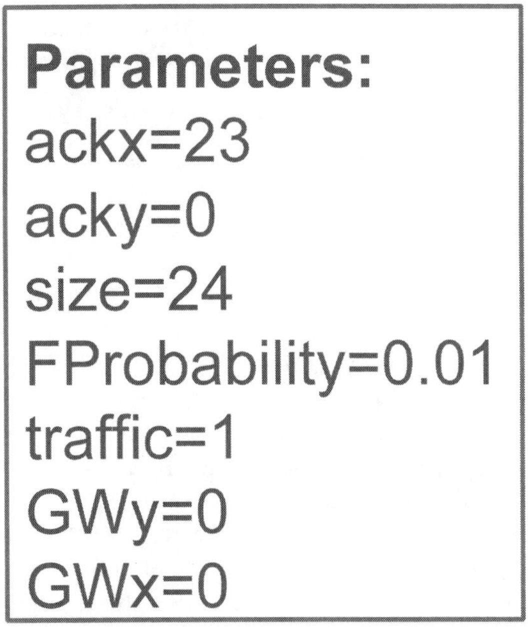

Parameters:
ackx=23
acky=0
size=24
FProbability=0.01
traffic=1
GWy=0
GWx=0

Figure 5.6 The Parameters used for representing some characteristics of the network for a simulation.

forming model checking experiments, the Statistical Model Checking (SMC) module of the UPPAAL tool was used. Statistical model checking allows for checking a statistical sample of the model's state space to deal with the state-space explosion problem. The main model checking experiment done was for checking that the design (parameters and routing algorithm) was deadlock-free in the presence of multiple packets being routed simultaneously. The query simply asked for the total number of acknowledgment packets after the end of the execution. The case where not all acknowledgment packets are received is an indicator for the presence of a deadlock. A second experiment queries the total clock time of an execution and it is used to measure performance. These properties are checked using the following form of SMC queries:

$$\text{E}[\leq \texttt{timesteps}; \texttt{times}](\texttt{property})$$

where timesteps is the number of time-steps each execution trace is run; times is the number of traces being analyzed and property is the property under investigation. Model checking results were an important tool in the process of developing the final HSF design. Through an iterative process, several designs of the HSF were proposed and were formally evaluated using the SMC. For example, in the case where a problem, e.g., a deadlock, was found using the SMC, a non-statistic model checking experiment was run that could generate the problematic trace and instance. Based on the trace analysis the design team could then suggest an alternative design by changing the parameters of the model. Formal evaluation as part of the design process complements the use of simulation testing software, since it can be easily deployed to exhaustively check all the possible execution interleaving. Model checking has led to the collection of valuable information about necessary and/or sufficient conditions for deadlocks in other topology configurations. This information can lead to a more efficient usage of more advanced configurations in the future [84].

The UPPAAL timed state machine of the switch controller is presented in Fig. 5.7. In addition, the functionality of the Gateway is described by a separate state machine. A set of state machines that interact using input/output actions constitute the model of the HSF. The controller state machine has two states (idle and processing) and can perform input/output actions on two input channels and two output channels. At the idle state, the controller waits for an interaction on one of the two input actions that indicated the receipt of a packet. Upon such an interaction, the machine proceeds to the processing state and at the same time receives and buffers the packet. At the processing state, there are several actions that might be enabled. Firstly, the

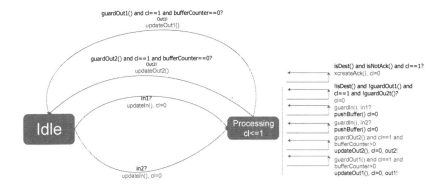

Figure 5.7 UPPAAL State Machine for the Switch Controller.

input actions are enabled in the case where there exists buffer space and more packets can be received. At the processing state, the first packet from the buffer is being inspected and if its routing target is a different controller then the corresponding output action is enabled following the routing algorithm. If the packet's routing target is the current controller, then the packet is consumed and an acknowledgment packet is prepared and routed accordingly. Each transition, if enabled, requires a clock tick to be completed. Clock ticks allow for counting time, thus having a metric for measuring performance. The model was designed so it can be adjusted for several parameters. The main parameters that interest the current evaluation are: the network size, the routing sequence of packets, the position of the Gateways, and the buffer size. The formal evaluation methodology together with the obtained results for particular scenarios appear in [84].

5.5.4   HyperSurface Emulator

This section presents the development of a hardware emulator for emulating HSF functional behavior. The emulator is able to emulate the HSF topology and the algorithmic design. The purpose of the HSF emulator is to facilitate the development of the network protocols required for the operation of the HSF System. Particularly, this initiative:

1. Emulates the intra-tile (ASIC chip) communication protocol, to test in advance and ensure its validity at a high level. The software scientists and networking experts current work focuses on

the evaluation of the proposed network controller mechanisms, using the various evaluation methodologies, which are being expanded by i) considering further application-specific metrics as needed within the context of the project, and ii) by taking into consideration measurements provided by the prototype chips that are currently under development (e.g., timing information) to produce more precise results.

2. Provides a layer of abstraction from the ASIC (switch fabric) prototype manufacturing progress, to enable the development of the HSF gateway hardware, firmware and software for this prototype in a disjoint manner.

Moreover, the emulator is used for the testing and evaluations of the modules to be integrated to the HSF prototype. More specifically the emulator will be used to:

1. Emulate the topology and the routing algorithms of the design and is used to test and debug the HSF down at the hardware level. Connect hardware elements so they can implement the topology and communication protocols of the HSF. Configuration packets, Acknowledgment packets, initialization packets, their format, and their way of transmission are as close as possible to the actual HSF implementation.

2. Complement the networking simulators and the formal evaluation models and can provide further results for analysis and evaluation. The emulator aims to implement the functionality requirements of the HSF design and provide measurements based on the metrics identified for the evaluation of the HSF. The advantage of the evaluation of HSF results comes from the fact that these are measured under more realistic hardware assumptions, and the interplay with the actual software implementation in the GW in contrast to abstracted software simulations and models.

3. Develop the hardware and the software for the GW and the User Device (UD). Given an HSF emulator that approaches real conditions, the GW, the UD and their interface can be developed and debugged using the emulator. Moreover, it can then be easily adapted in the actual HSF implementation with the minimal necessary modifications.

4. Implement and evaluate more sophisticated design alternatives of the HSF beyond the currently planned implementation. More sophisticated designs include more 'intelligent' routing that can

identify and tolerate faults. Experimenting with the emulator is useful for proposing future designs and implementations beyond the current HSF proposal. Importantly, one can also experiment with options such as wireless interconnectivity of the controller nodes, opening up a new direction of research in the networking aspect of the HSF.

The emulator is an experimental setup which aims to emulate the HSF device using a network of small computers namely, the Raspberry Pi. The purpose of this experimental setup is to implement an emulator so that the various modules (UD, GWs, controller nodes' network functionality, wired and wireless communication channels) are tested and evaluated. This step ensures that the selected modules will work according to the functional specifications of the HSF prototype. The design and development of the 'GW device' algorithms (routing and fault tolerance), GW and software on the user device will be done so that the compatibility between them and the final HSF prototype is ensured at the highest possible degree.

The aim is to speedily integrate the modules with the final HSF prototype. The successful integration may require some modifications with respect to the final functional and technical specifications of the prototype.

The emulator depicted in Fig. 5.8 consists of three devices, the UD (e.g., desktop and/or tablet), the GW and the HSF emulator. The UD will be used to send user directives (reference commands) to the GW which reads the necessary configuration commands from a database via the backend/cloud and sends them to the emulator via a 4-phase asynchronous communication interface. A wireless network of Bluetooth modules is used for the communication between the devices.

The emulator consists of a network of Raspberry Pi devices interconnected in order to emulate the nodes of the HSF. The design consists of a 6×6 nodes (36 in total) which are to be emulated on the Raspberry Pi network and are depicted in Fig. 5.9. There is a total of eighteen Raspberry Pi devices each one emulating two controller nodes. There are one GW and one Tile GW which are connected on each side of the emulator, a power supply capable of supplying all modules and a network switch is used for debugging/testing purposes. The GW consists of a Tile GW and the HSF Gateway. The HSF Gateway has the intelligent algorithms for routing/fault tolerance and the Tile GW is used for communication to the HSF emulator, error detection, data transmission, acknowledgments, etc. A photo of the emulator is shown

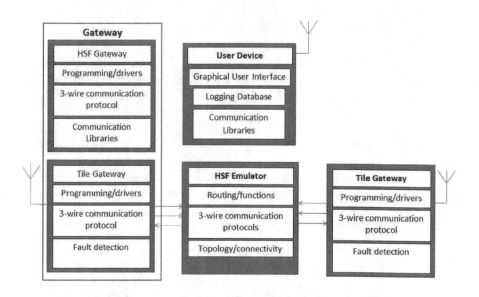

Figure 5.8 Block diagram of the HSF emulator.

in Fig. 5.10. The HSF emulator, the GW, Tile GW and power supply are clearly shown on the photo.

The functional block diagram in Fig. 5.11 describes the general functionalities required by the emulator. The User Device (UD) is a Tablet with an android operating system. A graphical user interface will be used to send user directives to the Raspberry Pi (sub-module of the HSF GW). The HSF GW will run all the smart algorithms necessary to configure and monitor the status of the HSF emulator. Two GWs (TGW1 and TGW2) are used to transfer the data from the RPi to the HSF emulator (configuration commands) and from the HSF emulator to the RPi (configuration commands acknowledgments). In particular the diagram shows the different software modules that implement the functionality of the HSF GW Interface. The arrows of the diagram show the information between the different software modules. The user provides directives through the Graphical User Interface (UI) on the User Device (UD). The directives are then being sent to the HSF GW Device through the Bluetooth Wireless Communication modules on the UD and the HSF GW Device, respectively. The software of the HSF GW device is executed through a number of threads:

- Main thread. Coordinates the functionality of the HSF GW Device.

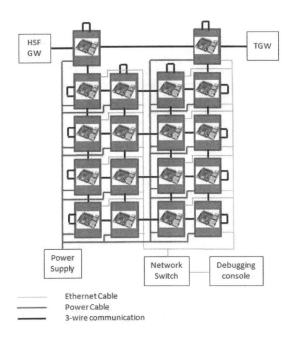

Figure 5.9 HyperSurface emulator with 36 nodes.

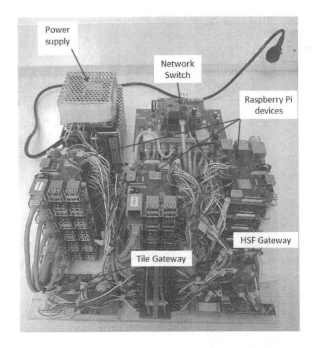

Figure 5.10 Photo of the developed HSF emulator.

- User directive thread. Receives directives from the user and creates a new event for the Main thread.

- Ack Thread. Registers the time packets that are being sent and expects their corresponding acknowledgments. In the case a timeout is detected the Ack Thread will create an AckTimeout event and send it to the Main Thread.

- GW Thread. Monitors the TGW1 for timeouts. In the case a timeout is detected it creates the corresponding timeout event and sends it to the Main Thread.

The information flow in the HSF GW Device is as follows: A configuration directive will first be stored on the configuration directive queue and subsequently received by the configuration user directive thread that will notify, through an event, the Main thread for a new directive. The Main Thread will then go into Directive Event Mode, it will read the directive and use the Server Interface Module to translate the directive into configuration values. The Main thread will then go into Send Packet Mode and through the Smart Programming module will create HSF packets and decide on the order of the sequence that is going to be sent. The Send Packet Mode will register packets, if needed, using a callback to the Ack Thread. It will then send the packet to the Gateway Communication Interface so it can be subsequently sent to TGW 1 via Bluetooth Protocol. If the Main Thread has no actions to perform, it waits for events at the Idle Mode. The Main thread also logs events on the logging Database through the Logging Database Interface. The Main Thread also prepares Status Messages that are sent to the User Device through the User Device Communication Interface. The TGW Communication Interface besides sending packets, receives the acknowledgments from TGW2 and Timeout Detection packets from TGW1. The former packets are being queued and received by the Ack Thread. The latter packets are being read by the Gateway Thread. The Ack Thread will confirm packet acknowledgment. In the case a timeout occurs, the Ack Thread will create an ack timeout event and send it to the Main Thread. The Main Thread will then go into the Timeout Handling Mode and through the Smart Programming module will take actions for fault identification and fault tolerance. The GW thread will also create a GW Timeout Event and will send it to the Main Thread that will also take action accordingly. TGW1 is responsible for receiving packets from the GW Communication Interface and for sending them via the 4-phase asynchronous protocol to the emulator. It also detects Gateway 4-phase asynchronous protocol timeouts and sends flags through the Gateway Communication Interface. Finally, TGW2 is responsible for receiving acknowledgment packets from the emulator, via the 4-phase asynchronous protocol, and sending them to the GW Communication Interface.

The classes that implement the software modules are: Main Thread, Acknowledgment Thread, Gateway Thread, Smart Programming. The execution loop of the Main thread is implemented in the method run in class Main. Method run iterates through the execution modes that are formed through the different events of the Gateway API. Acknowledgment class implements the loop that checks for acknowledgment packet timeouts, through method run. Method register, registers packets to be checked for the corresponding acknowledgment. Class GatewayTimeout implements the loop that checks for timeouts at TGW1, through the corresponding method run. Finally, class Smart Programming implements methods that given a set of configuration values, create a packet sequence for sending to the HSF emulator. The Smart-Programming class will be extended in future versions to add additional functionality for handling more complicated cases of HSF configuration scenarios (e.g., fault tolerance, optimized HSF configuration.

The class diagram is shown in Fig. 5.13. There are nine classes for the operation of the system. They represent the communication interfaces between the devices.

- UserDeviceComm class: It can execute three operations on the user device. It sends a directive, initiates request status of the HSF emulator and it receives the status for the HSF emulator both functions through the Raspberry Pi.

- RPiComm class: It is able to receive the directive from the user device and send the status to it when requested.

- Status class: It contains information for each controller in a grid form. It returns controller information (e.g., DAC values, fault status etc.) for each controller when requested.

- Info: This class contains the status of a single controller. In particular, the status information contains the DAC values, whether the controller was faulty and the controller's x-y values.

- GatewayComm class: Reponsible for sending packets to the HSF and receiving acknowledgments. The GatewayComm class is also responsible for handling timeouts by registering sent packets that request acknowledgment to the pending list and

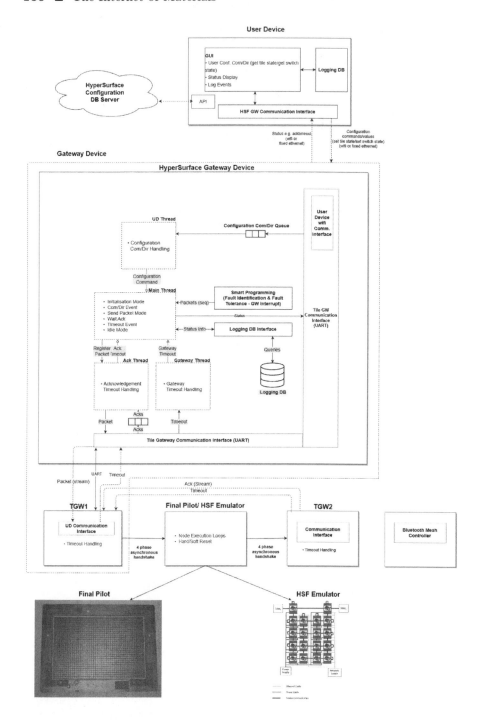

Figure 5.11 Functional Block Diagram of the HSF emulator.

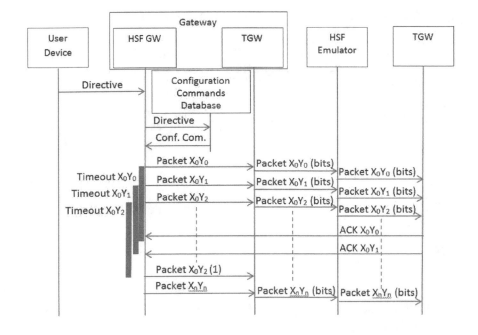

Figure 5.12 HSF emulator sequence diagram.

periodically checking whether the corresponding acknowledgments were received.

- Packet: The configuration command is formed here using the values for x, y and the rest of the configuration information.

- Directive: This class will include the directives that will be used for the formation of the configuration commands.

- Smart Programming class: Responsible for smart sequencing of the packets that will be sent towards the HSF. Smart sequencing depends on the existence of faults and the functionality/mode of operation of the gateway (e.g., normal, initialization, fault identification). The Smart Programming class has an internal Status state of the HSF.

- Event Handler class: Coordinates the operation of the gateway by acting on communication events and using callback methods on classes RPiComm and GatewayComm. It coordinates the functions of initialization, translation, sending configuration packets, sending status, and timeout/fault identification. This process is visualized in the sequence chart of Fig. 5.12.

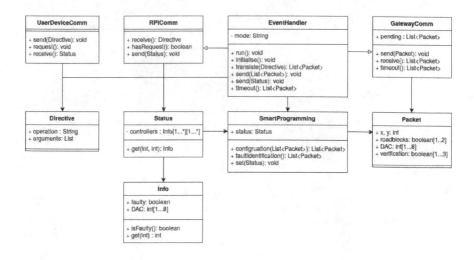

Figure 5.13 HSF Emulator Class diagram.

The use case diagram of the HSF emulator is depicted in Fig. 5.14. The involved entities are the following:

- The UD is the entity that can initiate directives through a Graphical User Interface (GUI). It is also possible to monitor the status of the HSF emulator.

- The GW receives the directives from the user device and creates a set of configuration commands. Then, it passes them to the HSF emulator using the 4-phase asynchronous communication protocol.

- The Tile GW receives acknowledgments from the HSF emulator and transmits them to the GW.

- The configuration commands database contains the allowed HyperSurface configuration commands for each of the predefined directives.

- The Interrupt handling manages interrupts initiated by errors (database, communication, or hardware) or data transmit communications originating from the intelligence device or the GWs.

Using the information returned by a successful HSF emulator initialization, a user can perform various tasks using the graphical user interface: i) set the directive for the HSF emulator, ii) get the status of the current HSF emulator configuration iii) request information

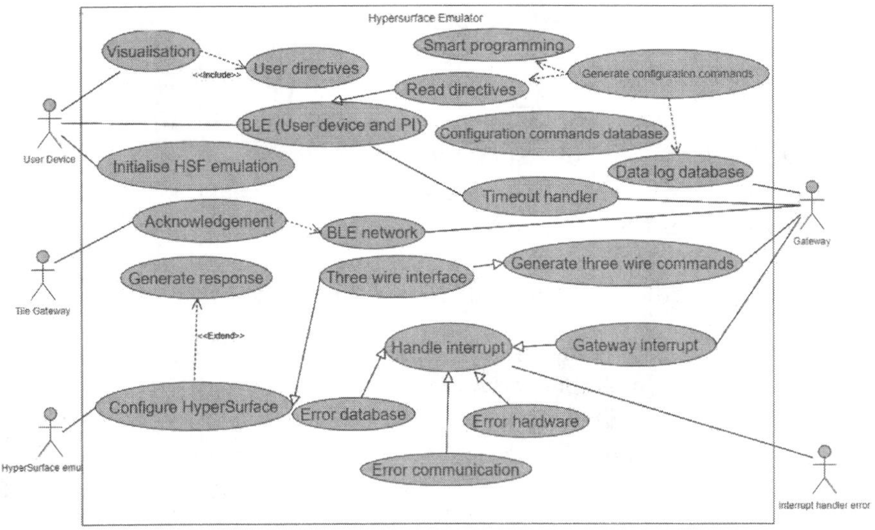

Figure 5.14 Emulator Use Case diagram.

about the 'health' status of the emulated HSF. The interrupt handling service is responsible for dispatching events like Database errors due to invalid data queries or null results, communication errors, due to network-related errors, hardware errors due to the detection of failing hardware components of the HSF emulator (emulating faults), Gateway interrupts raised by the HSF emulator GWs.

5.6 THE CONTROLLER-GATEWAY COMMUNICATION PERSPECTIVE

This section presents the HSF GW communication perspective and its modes of operation. On each mode of operation, controllers will be receiving and sending packets that contain information specific to the operation mode. Each operation mode will be interpreting the format of the packet following the needs of the functionality of the network.

5.6.1 Gateway Functionality

The GW basic operation is depicted in Fig. 5.15. It executes four main functionalities:

■ HSF configuration: This functionality receives a directive from the user and configures the HSF as desired. The directive is 'translated' into a set of configuration commands (one for each controller node) 'extracted' from a database, populated with the mappings between controller settings and desired user directives. The GW sends the configuration commands in an 'intelligent' customized sequence (designed to minimize delivery time and congestion) to the controllers and receives an acknowledgment for each one of them. In case no acknowledgment is received within an expected time frame, a timeout event is initiated and the Gateway goes into the fault identification mode.

■ Fault identification: This functionality is initiated after a time-out event. Timeout events occur in two cases: i) if the acknowl-edgment of a configuration command does not arrive within a predefined time window and ii) if a configuration command is 'stuck' at the GW (unable to move further into the network of controller nodes) within a predefined time window. This situa-tion is a consequence of a faulty node in the network: due to the asynchronous nature of the controller communication, the neigh-bours of a faulty node are unable to detect the failure of their neighbour, and their inability to forward packets results in their buffers becoming full with incoming packets, a phenomenon that propagates all the way to the GW. The fault identification algo-rithm then identifies the faulty control node and initiates the fault management function.

■ Fault management: This function takes remedial actions in the event that a faulty controller is identified. It forms a com-mand that includes bypass information necessary to controllers to avoid routing through faulty controllers that may lead to a non-functioning HSF.

■ HSF status: The user via the user device GUI can monitor the status of the HSF. The purpose of this function is to monitor and update the user about the status of the HSF.

The GW operates in an intelligent way to coordinate the function-ality of the HSF, given the conditions of operation, e.g., HSF config-uration time, presence of faults, etc. A state diagram that gives an overview of the GW intelligence is depicted in Fig. 5.16. The state dia-gram shows the events that are able to trigger the main functionalities of the GW. Three main states of operation ensure that the HSF has an interruptible operation:

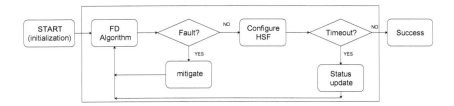

Figure 5.15 The gateway basic functionalities.

The "Normal/Fault Management" mode of operation is where the GW translates the directive of the user into a set of configuration commands (packets) that are intelligently sent into the HSF. This mode of operation assumes knowledge of faulty controllers, if any, and is responsible for routing configuration packets so that the faults are isolated/tolerated during the operation. Whenever a configuration packet reaches its target (controller to be configured) an acknowledgment towards the GW is issued. Acknowledgments provide knowledge for each configuration packet delivery to the GW and are an integral part for the successful coordination of the HSF. Whenever a packet is prepared, the GW proceeds to the "Ready to Send" state, where it deploys low-level communication mechanisms to send the packet to the HSF network. As long as packets are received by the network and successfully acknowledged without a timeout, the state remains in the "Normal/Fault Management" state. Timeouts are an integral part for proper coordination of the HSF functionality by the GW. An acknowledgment timeout can be triggered by the lack of an acknowledgment at the "Normal/Fault Management" state, where a separate timeout counter is monitored for each send configuration packet. A timeout can also be issued when a packet waits at the "Ready to Send" state, indicating that the HSF network cannot propagate packets into the network of controllers. The occurrence of timeouts indicates the presence of a fault in the delivery of a configuration packet to a specific controller or non-delivery of acknowledgment. In this case, the operation mode proceeds to the "Fault Detection" state. Note that even though the GW knows that there is a fault, it does not know which controller has become faulty (it could be any controller in the paths towards the delivery of the configuration or acknowledgment packets). The "Fault Detection state" is responsible to issue a reset signal globally to all the HSF controllers. Each controller will then perform a soft reset; it will delete all the packets from its buffers and return to an ini-

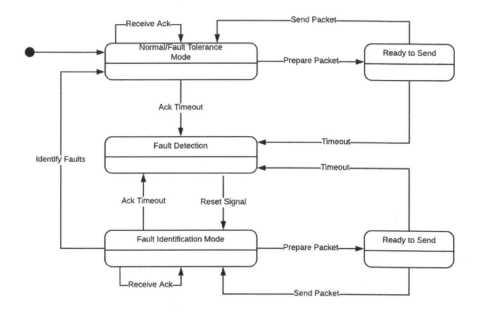

Figure 5.16 Gateway intelligence state machine.

tial state of execution. However, it will not erase its addressing data. After the reset signal is issued, the GW proceeds to the "Fault Identification Mode". In the "Fault Identification" state, the GW takes all the necessary actions to identify which HSF controllers are faulty. The algorithm sends packets, via the "Ready to Send" state, towards different controllers of the HSF and relies on the use of ready-to-send timeouts to calculate the address of a faulty controller. Thus, with the issue of a timeout the GW proceeds to the "Fault Detection" State, where a reset signal is issued and the state proceeds to "Fault Identification" in order to continue the operation of the fault identification algorithm. Once the fault identification algorithm terminates, the state returns to "Normal/Fault Management" mode with an updated knowledge on faulty nodes.

The normal mode flow chart is shown in Fig. 5.17 (left). A set of directives is stored on the user device where each directive represents a particular configuration of the HSF that the user may desire. The normal mode assumes knowledge of all the faulty controllers of the HSF. The user sends the desired directive from the graphical user interface installed on the user device to the Gateway where the configuration commands are formed and executed by the developed routing algorithm. Under normal operation (i.e., no new faults or timeouts are presented) the HSF control nodes return an acknowledgment for each successful execution of the configuration commands set.

The routing algorithm ensures proper configuration of each control node via the configuration commands. What happens if there is a fault on a controller, though? As mentioned above, when a timeout event occurs, the fault identification algorithm is initiated, which locates the faulty controller and takes remedial actions to ensure uninterruptable operation of the HSF. This is done using a set of configuration commands so that the configuration commands avoid using this 'faulty' control node for the routing. The flowchart of this algorithm is given in Fig. 5.17 (right).

Various types of information are communicated to the user about the status of the HSF through the graphical interface in order to keep the user updated about the status of the HSF. The flowchart in Fig. 5.18 shows the status algorithm, which is executed by the GW. The user can initiate a status request to update the interface or the interface is periodically updated by the GW. The status indicators include the following:

1. Fault alarm flag (timeout): This flag informs the user that there is a fault in the HSF.

2. HSF configuration command successful flag: When the user sends a directive to the Gateway, a flag is returned in case of success or failure.

3. Configuration command set formation successful: When the user sends a directive to the Gateway, the Gateway returns a flag in case of success or failure.

4. Mode of operation (normal, fault tolerance) fault detection, fault identification: The GW updates the user interface on the mode of operation of the HSF.

5. Performance restoration after fault event: The Gateway informs the user interface about the successful performance restoration after a fault event.

The software of the user device consists of a graphical user interface (GUI) and the libraries for wireless communication (Bluetooth in the pilot case) with the GW. The purpose of the GUI is to assist the user to send high-level directives to the GW. When a directive is sent to the GW, a set of configuration commands is prepared and passed to the HSF. The configuration commands are stored in a database which

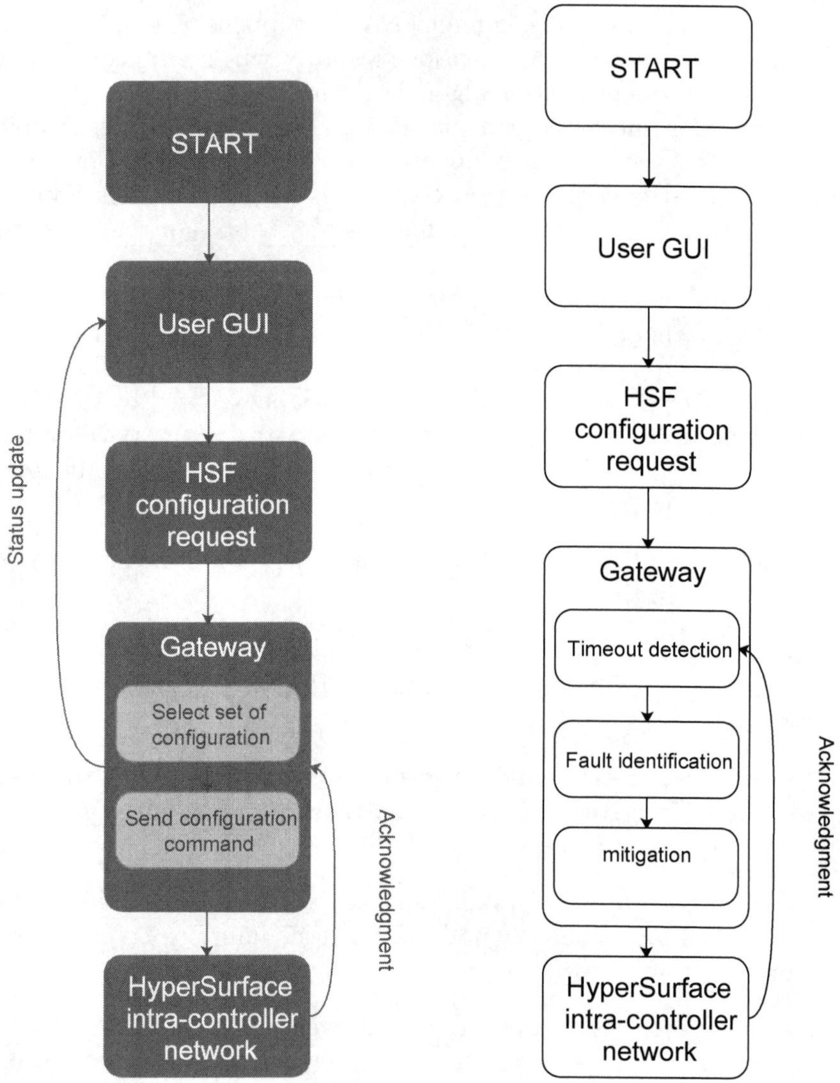

Figure 5.17 Algorithm flow for normal operation (left) and for fault-tolerance (right).

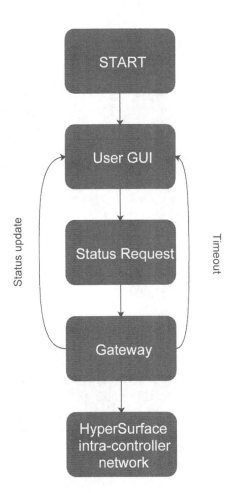

Figure 5.18 HyperSurface status update algorithm.

can be placed in the UD or in a cloud. Specifically, the graphical user interface aims to assist the user to: i) input and send the high-level directives to the HSF; ii) set the configuration parameters if needed; iii) obtain information about the status of the emulated HSF such as successful final configuration calculations, fault finding indications or any other flag necessary to inform the user about the status of the HSF emulator.

High-level directives are configuration commands which are sent from the user device in order to define the desired configuration of the HSF. As discussed earlier, the API callbacks include functions for setting the state of individual switch elements at each controller node, as well as retrieving their state. These callbacks have the general form:

$$\text{outcome} \longleftarrow \text{callback}(\text{actiontype}, \text{parameters})$$

where the action type is an identifier denoting the intended function, e.g., STEER, ABSORB, POLARIZE, and FILTER. Each action type is associated with a set of parameters. For instance,

1. STEER commands require:

    an incident wave direction, I,

    an intended reflection direction, O,

    applicable wave frequency, F.

2. ABSORB commands require no O parameter.

The Caller executes an EM Function deployment Callback function, which in turn invokes the Configuration resolver. The resolver queries the Configuration database and returns one or more configurations that are combined into the configuration that best matches the intended EM function. The configuration is conveyed to the GW using the corresponding protocol. The GW reformats the received configuration and executes actions to inform (and thereby set) each separate controller node switch element accordingly. The HSF controller communication protocol is employed towards this end. Finally, each switch element is set to its intended value, thereby "consuming" the corresponding part of the configuration.

The user directives express the desired behavior of the HSF and are 'translated' into a set of configuration commands using compilation software. The compilation of the set of directive commands is done offline and the results of the compilation are stored in a database either locally on the user device or on the cloud. The user device can connect to the database and retrieve the required compilation result.

A wireless communication network will be used for the data transfer between the User Device, the GWs and the emulated HSF. A mesh type Bluetooth device will be used for the implementation of the wireless network. The communications network comprises:

- Wireless link between the tablet and the GWs all in a mesh type of connection.

- Wireless link between GW and Tile GW using the Bluetooth mesh protocol.

- Synchronization of the data between the GWs.

The communication between the GW and the HSF tile is done using a 4-phase asynchronous communication interface described in Section 5.5.4.

The synchronization in the wireless network is critical for the uninterruptable communication between the devices as well as the data flow in the HSF tile. The communication between the devices is given in the UML diagrams in this section. Data congestion avoidance mechanisms are implemented during data traffic.

5.6.1.1 Software/Firmware Design and Development

This section gives a description of the firmware development on the Tile GWs, the software development on the HSF GW and the User Device for the HSF emulator. The GW consists of the HSF GW (Raspberry Pi device) and the Tile GW (custom development). Figure 5.19 shows the block diagram of the communication between different elements. Figure 5.20 illustrates the development setup. The HSF GW software developed is running on a small size and low power consumption device called Raspberry Pi Zero shown in Fig. 5.21. In the next sections, the firmware developed for the communication between HSF GW and UD is described. The communication between the Bluetooth and RPi that forms the HSF GW is done using the UART port. The flow of the data from the UART to Bluetooth device (CSR1011) is received using the UART firmware and managed with a librUOBQueue library including the queuing of the data. The UART to Bluetooth queue, is responsible for managing the data that is received through the UART protocol. In addition, this library checks that the data it is valid. If the data is for example a configuration command it is stored in the queuing buffer. Each configuration command is 16 bytes with two additional flag bits for indicating packet availability and one acknowledgment bit

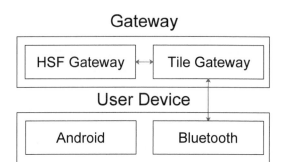

Figure 5.19 Gateway submodules communication and communication with the user device.

Figure 5.20 HyperSurface Gateway and user device development.

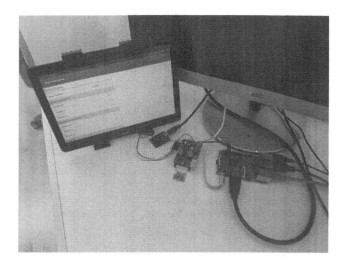

Figure 5.21 HSF gateway based on Raspberry Pi zero (65cm x 30 cm).

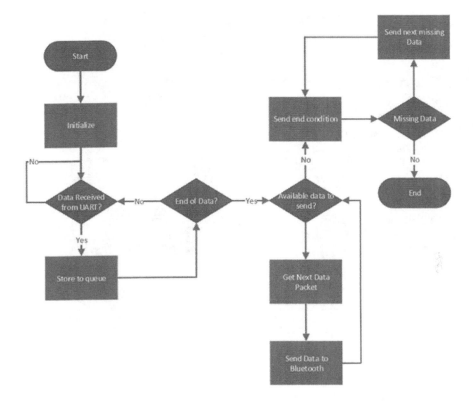

Figure 5.22 Flowchart for the Bluetooth device UART firmware.

(this bit indicates whether the configuration command has arrived at the destination node of the HSF). If there are packages that are not acknowledged then there are functions that collect those and pass them for further processing i.e., resend them in the mesh network. Extra functionalities have been implemented to manage the unacknowledged packets more efficiently which means improved reliability and speed of the data transfer. This buffer queue is implemented on the HSF GW side. The detailed flow of the data is shown in the flowchart of Fig. 5.22.

5.6.1.2 Tile Gateway Communication Interface Firmware

This section describes the implementation of the interface on the Tile GW (receiver and transmitter sides) that is used to pass and receive the configuration commands from the HSF Tile. The communication protocol is based on the 4-phase dual rail protocol that is used in two cases: i) when the Tile GW acts as a receiver and the Tile acts as

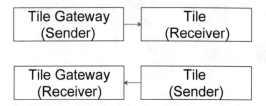

Figure 5.23 3-wire communication between TGW and Tile.

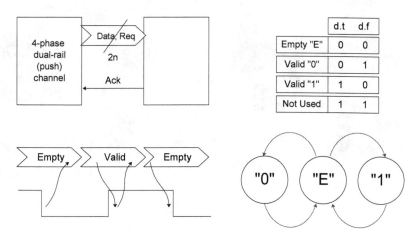

Figure 5.24 The 4-phase dual rail protocol: top left: communication channel, top right: truth table for data validity check at the receiver side, bottom left: signal transition and bottom right: state transition.

transmitter and ii) when the Tile acts as a receiver and the Tile GW acts as a transmitter (see Fig. 5.23).

In Fig. 5.24, a graphical interpretation of the protocol is illustrated. The request signal is encoded into the data signals. The data use two wires per bit. One wire is d.t and used for signal logic 1 (true) and the other wire is d.f and used for signal logic 0 (false). The request signal is set to HIGH only when the all data signals are valid. This happens when each bit has only one of its two signals (d.t or d.f)[1] HIGH (logic 1).

As an example, consider the following: The word information that the sender sends to the receiver is the binary number $(100)_2$. For each bit, we have that:

$d.t_1 = 0, d.f_1 = 1$
$d.t_2 = 0, d.f_2 = 1$
$d.t_3 = 1, d.f_3 = 0$

---

[1]d.t: data true, d.f: data false.

The data shown above is valid since each bit has one of the two signals HIGH and one LOW. This means the request signal will be set to HIGH. An important observation at this point is that the request will never be set to HIGH before the data is valid at the receiver, because the data signals become valid first and the request signal is enabled afterwards. Once the receiver takes the data, it acknowledges the sender by setting acknowledgment to HIGH. The acknowledgment at the sender's side makes the data invalid and, thus the request signal is set to LOW. The acknowledgment signal is set to LOW and concludes one cycle. The communication cycle between a sender and a receiver is done as follows: First, the sender sets both d.t and d.f at the receiver input. According to the validity of the bits, the receiver does the following:

- Case 00: The receiver does not change its state because the data is empty.

- Case 01 or 10: The receiver receives the data and once done, it acknowledges the sender by setting the acknowledgment signal HIGH. The receiver then waits for the data to become invalid to lower the acknowledgment signal. On the other hand, the sender, once it sees a HIGH acknowledgment signal from the receiver, it lowers both input bits.

- Case 11: This case is abnormal and should not happen. If for some reason this occurs the behavior of the protocol cannot be predicted.

Next, the firmware library for implementing the 3-wire communication protocol is described which is used in both send and receiving Tile GWs. Figure 5.25 shows the flowchart for the protocol operation. The protocol is able to work as read or write without any interference from the user. A timeout function is also implemented which helps the protocol to unstack in case there is some problem with data sending/receive. The timeout is configurable, i.e., to change the timeout duration.

An algorithm to manage multiple channels (in the case of the tile gateway we have 2 input and 2 output channels) has been developed. Here the description of this algorithm is done with the help of state machine diagrams. The diagram in Fig. 5.26 shows the operation of the algorithm. The following states describe the flow of the algorithm:

1 The first (initial) state is called when one or more channels require action.

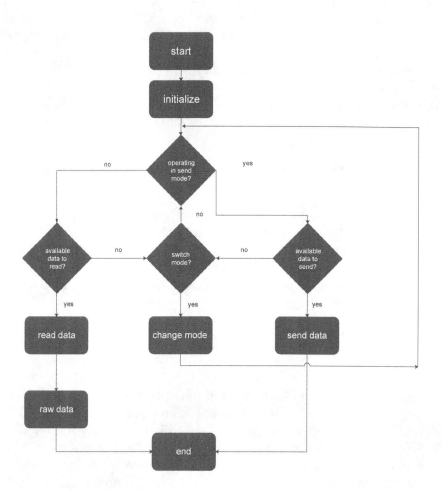

Figure 5.25 Flowchart for tile gateway sender/reader modes.

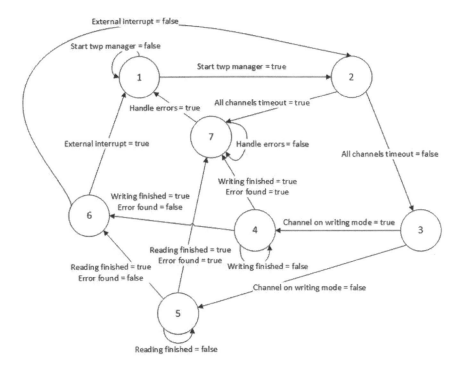

Figure 5.26 State machine of the tile gateway for managing multiple channels (2 input and 2 output channels).

2 The second state checks if all the channels are inactive or on timeout, if not proceeds to the next active channel.

3 The third state checks if the selected channel is on writing or reading mode.

4 This step is another state machine in this algorithm for the writing process which is described below.

5 State machine for reading process, refers to reading state machine.

6 Checks if there is an external factor that needs attention, and interrupts the manager, if not, returns to state 2 and gets the next active channel.

7 If an error occurs, handle the error and stop the process.

5.6.1.3 Error/Fault Detection

The algorithm is able to detect error/fault during the writing/reading process. The error detection during the writing process of the three wire protocol is described with the state machine diagram in Fig. 5.27. The algorithm step description is as follows:

1 When the write process is called, the first state is used to initialize the required variables.

2 The second state, reads the next bit from the communication data buffer.

3 The third state checks whether the ACK=false. If this condition is not met the timeout occurs (timeout=true) and jumps to the error handling state, i.e., state 8.

4 If ACK=false and timeout=false, the fourth state sets the DATA true and DATA false accordingly. When set data lines=true the state machine moves to state 5.

5 At this state the algorithm is waiting for the ACK to become true (which means that the ASIC chip successfully received the data), if this condition is not met the timeout occurs and the algorithm jumps to state 8 (error handling).

6 This state sets both data lines to the false state. When set data lines are low=true the state machine moves to the next state.

7 When ACK=false, this means that the ASIC chip changes the state of the ACK line. If the condition is not met a timeout occurs and jumps to the error handling state (i.e., 8). In addition if there are more bits to be sent the machine returns to state 2, otherwise it jumps to state 1 waiting for the next data to be sent.

8 Error handling routine if timeout occurs during the reading process.

The error detection during the reading process of the three wire protocol is described here with the state machine diagram in Fig. 5.28. The algorithm state description is as follows:

1 The initial state, used to initialize the required variable before the reading process begins.

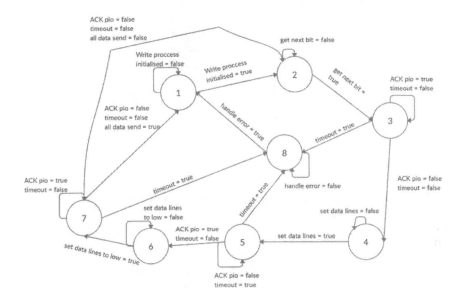

Figure 5.27 Error detection state machine during the writing process.

2 Waiting for the condition DATA true to be unequal to DATA false in order to continue; if this condition is not met after a certain time, a timeout occurs and jumps to the state 7 (error handling).

3 This state stores the received bit to the communication data buffer.

4 Waiting for the condition DATA true OR DATA false == false, if the condition is not met after a certain time the timeout occurs and the algorithm jumps to the error handling state.

5 If there are more bits to read, return to state 2; if all the bits have been read, continue to state 6.

6 Store all the data to the external buffer.

7 This is the error handling routine initiated on timeout during the reading process.

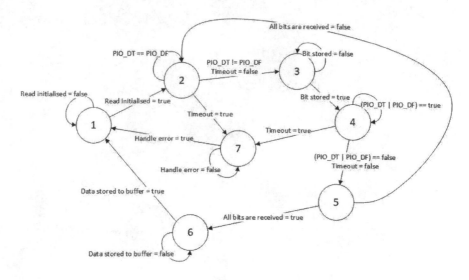

Figure 5.28 Error detection state machine during the reading process.

5.6.1.4 Bluetooth Mesh Firmware

A wireless communication network will be used for the data transfer between the User Device, the Gateway and the emulated HSF. The communications network comprises:

1 a link between the user device and the GW,

2 a link between the GW and the HSF emulator,

3 a link between GW and Tile GW,

4 synchronization of the data between the Tile Gateways.

A mesh type Bluetooth device will be used for the implementation of the wireless network, based on the CSR1011 Qualcomm chip that will be used on the Tile GW. The synchronization in the wireless network is critical for the uninterruptable communication between the devices as well as the data flow in the HSF emulator. Data congestion must be avoided during the data traffic. The diagram of Fig. 5.29 shows a typical mesh configuration using Bluetooth devices for controlling multiple HSF tiles.

In order to achieve the highest possible data transfer rate to the Bluetooth mesh network but also reduce at the lowest possible rate the collision rate of packets, the receive duty cycle of the GWs' Bluetooth chips is configured on the fly accordingly. Before the HSF GW starts

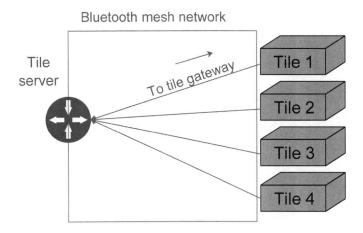

Figure 5.29 Bluetooth mesh network for a group of the hypersurfaces (tiles).

sending the data packages to the tiles' GWs, it sends a specific package with which it configures them in "receiving mode". Then when all the data is transmitted, another configuration packet is sent to set the GWs to a slower receive duty cycle. In addition, when the Tile GW1s receives the last packet from the HSF GW, a full check of the received data is performed. If there is a missing package on one of the Tile GW1s, a relative message is sent back to the HSF GW, containing also the ID of the Tile GW1 that sent it. The process is repeated until all the tile gateways receive all the data. During the whole process the Tile GW2 sends any acknowledgment packets, gathered from the tile, to the HSF GW.

The state machine of Fig. 5.30 describes the HSF GW mesh network operation. As shown in the figure the Gateway can be at one of the following states:

1 State where the HSF GW is waiting to connect with the UD.

2 When the device connects with the UD, it broadcasts the devices' IDs to the GWs.

3 Waits until data needs to be sent to the GWs.

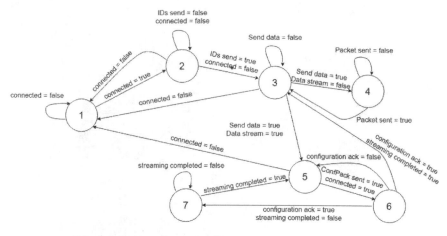

Figure 5.30 Mesh bluetooth state diagram.

4 Broadcasts a single packet into the network, no need to make any adjustments for performance.

5 Sends the configuration packet, to configure the receiving duty cycle of the Bluetooth GWs.

6 State that checks if the configuration packet has been acknowledged by all the GWs.

7 Streams the data packets into the Bluetooth network.

5.7   THE HYPERSURFACE WITHIN CONTROL LOOPS

The design of the controller network topology and routing protocols may have a direct impact on the performance of the HSF. Delay-sensitive applications for instance, can be affected by the communication delay due to the controller network. Such an application is the Beam steering application, envisioned as a solution to the non-line-of-sight connection required by the 5G communications technology. The mmWaves are known to be easily obstructed by various obstacles, from buildings in outdoor areas to furniture pieces indoors. Thus, the development of a method to overcome signal blockage is crucial for 5G technologies [136]. Different types of reflecting technologies have been proposed in the literature to tackle this problem such as the works in [200], [17] and [157]. Recently the HSF paradigm has been proposed to manipulate the e-m waves, allowing anomalous reflection, full e-m absorption, refraction, and polarization control.

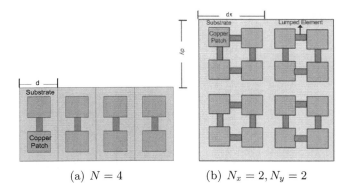

(a) $N = 4$ (b) $N_x = 2, N_y = 2$

Figure 5.31   Illustrative supercell structure.

In this section we present work in progress regarding the employment of feedback in the design of a beam steering application using the HSF paradigm. The aim of the control mechanism is to maximize the received signal strength (RSS) at the receiver through redirecting the incident wave on the surface towards the receiver. A system of a transmitter, a receiver and the HSF is considered.

We implement the use of a control algorithm that regulates the angle of the reflected wave such that the maximum possible SNR is achieved at the destination.

In addition, we lay out a theoretical model for the control mechanism which is then validated through simulations. The significance of the work lies in posing a metasurface-based beamsteering problem as a feedback control problem, which has not been considered in the literature before.

5.7.1 System Model

Anomalous reflection, the functionality used for beam steering, is accomplished through the collective behavior of a number of unit cells combined to form a *supercell* as depicted in Fig. 5.31(a). The number of unit cells in a supercell, N, is used to describe the behavior of the reflection angle in the equation below [36]:

$$\frac{2\pi}{\lambda_0} \sin \theta_r - \frac{2\pi}{\lambda_0} \sin \theta_i = \frac{n2\pi}{Nd}, \tag{5.1}$$

where λ_0 is the incident wavelength, θ_r is the reflection angle, θ_i is the incidence angle, n is the diffraction order and d is the length of the cell. Thus, the value of the reflection angle relies on both the incident wave and the parameters of the metasurface. The considered metasurface supports the adjustment of N through software directives via the gateway. Therefore, N is selected as the control variable: by changing the total size of the supercell, a phase gradient is obtained across it via Eq. (5.1), which can be tuned to match the required reflection angle θ_r. The change in the gradient is achieved through varying the value of the RC load connecting the copper patches of the unit cell. In reality, N can only obtain integer values. However, in order to simplify analysis we assume continuous values for all variables. This assumption also allows us to obtain useful insights in this novel control application. We will leave the integer value case as near future work, which can be cast in the framework of switched systems, as well as the fine tuning of the RC load.

The metasurface considered in this analysis is planar, lying in the $xy-$plane (normal to $z-$axis), and consists of a rectangular grid of unit-cells arranged parallel to the x and $y-$axes, as shown in Fig. 5.31b. The dimensions of each unit-cell in the $xy-$plane are d_x and d_y, and the metasurface contains N_x by N_y unit-cells such that the metasurface contains a $N_x * N_y$ grid of unit-cells. Following Eq. (5.1), the governing equation for the diffraction mode elevation angle (i.e., the angle inside the plane of incidence) in the $x-$direction is expressed as:

$$\theta_{r,n} = \sin^{-1}\left(\frac{n\lambda_0}{N_x d_x} + \sin\theta_{i,n}\right) \tag{5.2}$$

Equivalently, diffraction modes arise in the $y-$direction according to:

$$\theta_{r,m} = \sin^{-1}\left(\frac{m\lambda_0}{N_y d_y} + \sin\theta_{i,m}\right) \tag{5.3}$$

where n and m are the diffraction orders related to x and y directions respectively, $\theta_{i,n}$ and $\theta_{i,m}$ are the incidence angles of x and y components of the incidence signal, and $\theta_{r,n}$ and $\theta_{r,m}$ are the corresponding angles of reflection of x and y components of the reflected signal as shown in Fig. 5.32b.

5.7.2 The Considered Model

Figure 5.32 shows the considered system model, where the aim is to direct the reflected wave towards a target point, i.e., the receiver location, through appropriate adjustments to the reflection angle $\theta_r = [\theta_{r,n} \quad \theta_{r,m}]^T$. The value of these adjustments can be found by estimating the distance between the current arrival location of the reflected

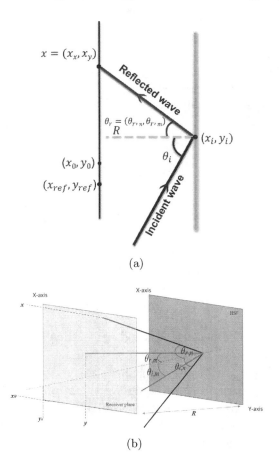

Figure 5.32 Illustrative reflection example.

signal and the desired target location denoted by $\tilde{x} = \begin{bmatrix} \tilde{x}_x & \tilde{x}_y \end{bmatrix}^T$, where $\tilde{x}_x(t) = x_x - x_0$ and $\tilde{x}_y(t) = x_y - x_0$. The incidence angle is assumed to be measured directly by the surface, whereas the RSS measurements at the receiver location are assumed to be fed to the system through the gateway. In Fig. 5.32, (x_0, y_0), (x_x, x_y), (x_{ref}, y_{ref}), (x_i, y_i) and R denote the target location, current location, a reference location, the incidence location and the horizontal distance between the reflective plane and the receiver, respectively. The following relationships can be deduced from the figure:

$$x = (x_x, x_y) = \begin{bmatrix} x_i + R \tan \theta_{r,n}(t) \\ y_i + R \tan \theta_{r,m}(t) \end{bmatrix} \tag{5.4}$$

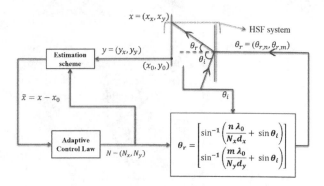

Figure 5.33   Closed Loop Feedback System.

By subtracting (x_0, y_0) from both sides of (3) we get:

$$\tilde{x}(t) = \left[\begin{array}{c} x_i - x_0 + R\tan\theta_{r,n}(t) \\ y_i - y_0 + R\tan\theta_{r,m}(t) \end{array} \right] \tag{5.5}$$

In the considered beam-steering application the objective is to adjust the value of $\theta_r = [\theta_{r,n} \quad \theta_{r,m}]^T$ through changing the size of the super-cell, $N = [N_x \quad N_y]^T$, using an appropriate control algorithm such that the error $\tilde{x} = (\tilde{x}_x, \tilde{x}_y) = 0$. The design of the appropriate control technique is, however, challenged by two problems. The one is the degree of uncertainty in the assumed model. Inaccuracies in underlying processes such as erroneous measurements resulting from the misplacement, or tilting of the surface for example, can be a source of uncertainty in the mathematical representations of these processes, e.g., Eq. (5.1). This uncertainty, renders open loop approaches, which use the underlying geometry of Fig. 5.32 to determine N based on the estimates of $\tilde{x}$, potentially unsuitable. Open loop approaches are less robust against system errors and uncertainty and this calls for the use of feedback which we consider and promote in this section. The second challenge is that the control algorithm is based on $\tilde{x}$ which is time varying and unknown. $\tilde{x}$ is found through the measurements of the RSS at the receiver. The relationship between the two is also subject to uncertainties. Therefore, a robust mechanism to estimate $\tilde{x}$ is crucial. As such, we consider the feedback based architecture of Fig. 5.33. The RSS and the current control input are fed into the estimation algorithm which generates estimates of $\tilde{x}$. The latter is then used by the control algorithm to update N. Both the control algorithm and the estimation algorithm take the form of iterative algorithms.

5.7.3 Control Algorithm

We derive our control algorithm based on the model described in the previous section. Since the reflection angle is dependent on the size of the supercell, the following nonlinear control law is proposed to update N.

$$
\dot{N} = \begin{bmatrix} \dot{N}_x \\ \dot{N}_y \end{bmatrix} = \gamma \begin{bmatrix} \tilde{x} N_x{}^2 d_x \sqrt{1 - (\frac{n\lambda_0}{N_x d_x} + \sin\theta_{i,n})^2} \\ \tilde{y} N_y{}^2 d_y \sqrt{1 - (\frac{m\lambda_0}{N_y d_y} + \sin\theta_{i,m})^2} \\ N_x(0) = N_{x(0)}, N_y(0) = N_{y(0)} \end{bmatrix}, \tag{5.6}
$$

where $\gamma > 0$ is a design parameter which affects the convergence properties of the algorithm. We then use Lyapunov arguments to establish the stability of the algorithm. In order to ensure stability the derivative of the selected Lyapunov function must be a negative definite (i.e., ≤ 0) [76]. The derivation details, for the interested reader, are in [21].

5.7.4 Estimation Algorithm

The authors in [126] show that an Extended Kalman Filter (EKF) [23] model can be employed to estimate the location of a beam. Here, we utilize the same mechanism for the prediction of (x_x, x_y), i.e., the coordinates of arrival of the reflected wave. Figure 5.33 depicts the estimation algorithm where the measured power intensity at the current arrival location and the number of unit cells in a supercell, N, are fed to the estimation scheme. The current location coordinates estimated by the algorithm are then subtracted from the receiver location coordinates to find $(\tilde{x}_x, \tilde{x}_y)$. The nonlinear system to be employed in the design of the estimator can be described by the following model:

$$
\tilde{x}_{(k+1)} = A\tilde{x}_{(k)} + B_w w(k) + B_u u(k) \tag{5.7}
$$

$$
z_{(k)} = h(\tilde{x}_{x(k)}, \tilde{x}_{y(k)}) + v_{(k)} \tag{5.8}
$$

where $h(\tilde{x}_{(k)}) = h(\tilde{x}_{x(k)}, \tilde{x}_{y(k)})$ is defined by the following quadratic function [126]:

$$
\begin{aligned}
h(\tilde{x}_{x(k)}, \tilde{x}_{y(k)}) = a_1 \tilde{x}^2_{x(k)} + a_2 \tilde{x}^2_{y(k)} + a_{12} \tilde{x}_{x(k)} \tilde{x}_{y(k)} \\
+ b_1 \tilde{x}_{x(k)} + b_2 \tilde{x}_{y(k)} + c
\end{aligned} \tag{5.9}
$$

and $z_{(k)}$ denotes the predicted power. $w_{(k)}$ and $v_{(k)}$ denote the process noise and the measurement noise, respectively. The nonlinearity of the system model explains the rationale behind using an EKF to estimate $\tilde{x}$. The algorithm of the EKF can be summarized as follows:

■ Prediction:

$$\tilde{x}_{(k+1|k)} = A\tilde{x}_{(k|k)} + B_u u(k) \tag{5.10}$$

$$P_{(k+1|k)} = AP_{(k)}A^T + Q_{(k)} \tag{5.11}$$

$$K_{(k+1)} = P_{(k+1|k)}H_{(k+1)}^T(H_{(k+1)}P_{(k+1|k)}H_{(k+1)}^T + S_{(k)})^{-1} \tag{5.12}$$

It is important to note that an additional term $(B_u u(k))$ appears in the prediction of system state as shown in (5.10), because (5.7) depends on $u(k)$, where $u(k)$ is known by the estimator. P is the predicted covariance matrix, $S = \sigma_v^2$ is the measurement noise covariance, Q is the process noise covariance,

$$Q = \begin{bmatrix} \sigma_w^2 & 0 \\ 0 & \sigma_w^2 \end{bmatrix} \tag{5.13}$$

A is a design parameter. K is the filter gain which is calculated based on the Jacobian of z represented by H_{k+1}:

$$H_{k+1} = [2a_1\tilde{x}_{x(k+1|k)} + a_{12}\tilde{x}_{y(k+1|k)} + b_1 \quad 2a_2\tilde{x}_{y(k+1|k)} \\ +a_{12}\tilde{x}_{x(k+1|k)} + b_2] \tag{5.14}$$

■ Update:

$$\tilde{x}_{(k+1)} = \tilde{x}_{(k+1|k)} + K_{(k+1)}[y_{(k+1)} - h(\tilde{x}_{(k+1)})] \tag{5.15}$$

$$P_{(k+1)} = (I - K_{(k+1)}H_{(k+1)})P_{(k+1|k)}, \tag{5.16}$$

where y is the measured power, h is the power predicted based on $\tilde{x}_{(k+1|k)}$ and I is the identity matrix, which is defined as follows:

$$I = \begin{bmatrix} 1 & 0 \\ 0 & 1 \end{bmatrix} \tag{5.17}$$

Remarks: It is to be noted that we have considered the additive noise in our problem. However, by considering the non-additive noise, Eq. (5.10) modifies to:

$$K_{(k+1)} = P_{(k+1|k)}H_{(k+1)}^T(H_{(k+1)}P_{(k+1|k)}H_{(k+1)}^T \\ +V_{(k)}S_{(k)}V_{(k)}^T)^{-1} \tag{5.18}$$

where Eq. (5.18) necessitates the computation of an additional Jacobian matrix related to $v_{(k)}$, which is defined as:

$$V = J_v = \begin{bmatrix} \frac{\partial h(\tilde{x}_{x(k)}, \tilde{x}_{y(k)})}{\partial v_{x(k)}} & \frac{\partial h(\tilde{x}_{x(k)}, \tilde{x}_{y(k)})}{\partial v_{y(k)}} \end{bmatrix}, (v_{x(k)}, v_{y(k)} \neq 0) \tag{5.19}$$

Parameter	Value	Parameter	Value
θ_i	0	λ_0	0.06 (m)
R	4 (m)	$d_x = d_y$	0.00912 (m)
$x_{ref} = y_{ref}$	0	σ_w^2	0.001
$n = m$	1	σ_v^2	0.01

Table 5.3 Parameters Values.

5.7.5 Performance Evaluation

In the previous section we have established the convergence properties of the system and proposed an EKF-based estimation scheme to
predict the current location of the beam. In this section we assess the
performance of the proposed models through simulation experiments.
The feedback system depicted in Fig. 5.33 has been implemented on
MATLAB SIMULINK. We assume that the incident signal is normal
on the surface, i.e., $\theta_i = 0$, the receiver is located at the reference point
such that $(x_0, y_0) = (x_{ref}, y_{ref})$. Moreover, the receiver is assumed to
be 4 meters away from the surface. The initial value of N_x and N_x is set
to 10 unit cells whereas the values of the rest of the system parameters
considered in the experiment are indicated in Table 5.3. It is important
to note that the analysis conducted in this work applies to the far field
(Fraunhofer) region. We assume that the transmissions take place at
the 60 GHz mmWave band, where the distance is typically between
tens of centimeters and a maximum of a few meters, depending on the
size of the antenna.

First, we investigate the convergence behavior of the system. Figure 5.34 depicts the rapid convergence of error close to zero for both
the x and y directions, where the x direction corresponds to the distance of the beam from its target in the $x-$axis (i.e., this distance is
controlled by changing the number of unit cells N_x) and the y direction
represents the distance of the beam from its target in the $y-$axis. It
can be observed from Fig. 5.34a and Fig. 5.34b that estimated position
of the beam shows a good approximation of the real position indicating
small estimation error.

We have established the convergence properties of the proposed
scheme using analysis and evaluated the performance through simulation experiments. Our results indicate that the proposed algorithm is
able to track the error in the beam location close to zero. The proposed
methods are based on assumptions that we aim to relax in the future.

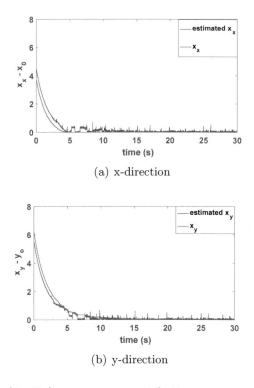

(a) x-direction

(b) y-direction

Figure 5.34 $\tilde{x} = (\tilde{x}_x, \tilde{x}_y)$ convergence with time.

In the future, we plan to consider a case of multiple beams and N will only assume integer values.

5.8 SUMMARY

This chapter presents a detailed overview of the design of the HSF networking aspect. In particular, it presents the HSF-CN Manhattan-like topology and the adopted communication protocols. The chapter also presents the design and the implementation of the HSF-CN Gateway as a core component that coordinates and configures the functions of the HSF-CN. The HSF-CN uses an adapted version of the XY routing protocol as a protocol which is simple enough to be easily implemented.

The design has undergone an extensive evaluation, using complementary approaches and techniques. In particular simulations and model checking techniques were used to extract information about the correctness of the proposed network protocols and analyze basic network metrics.

The evaluation of the design was also done with the implementation of an HSF emulator, which was very useful in testing lower level communication protocols and routines. The HSF emulator was also used to develop, test, and integrate the HSF-CN Gateway component.

The GW's main role is to connect the user with the HSF-CN. The GW is responsible for issuing software defined directives towards the HSF-CN in the form of configuration packets and it needs to be aware of the HSF-CN state at any point. The GW sends configuration packets and receives acknowledgment packets. Using timeout events the GW can detect the existence of errors/faults and can initiate and coordinate fault tolerance routines. The development of the Gateway was done on the HSF emulator. The fact that the emulator is closer to the pilot HSF prototype enables easy integration of the Gateway with the HSF prototype.

The chapter concludes with an exemplary application of the HSF for beam steering. An HSF can be used to steer EM beams for non line of sight connection. Beam steering uses a control loop mechanism to maximize the received signal strength at the receiver.

ACKNOWLEDGMENTS

This work was funded by the European Union's Horizon 2020 research and innovation program, Future and Emerging Topics (FET Open) under grant agreement No 736876. Aspects of this work build on VISORSURF, and the contribution of our colleagues is acknowledged.

CHAPTER 6

Internet of Things-Compliant Platforms for Inter-Networking Metamaterials

Andreas Miaoudakis

Foundation for Research and Technology Hellas, 71110, Heraklion, Crete, Greece

CONTENTS

6.1	Overview	....	200
6.2	Hardware Actuation Approaches	....	201
	6.2.1	RF Switching Elements	201
		6.2.1.1	PIN Diodes	201
	6.2.2	Controller to PIN Interface	202
		6.2.2.1	DAC	203
6.3	Controller Communication	....	205
	6.3.1	Controller to Controller Communication	206
		6.3.1.1	SPI	206
		6.3.1.2	I2C	207
		6.3.1.3	UART	207
		6.3.1.4	CAN	207
	6.3.2	Controller to Server Communication	208
		6.3.2.1	Bluetooth	208
		6.3.2.2	802.15.4	208
		6.3.2.3	Zigbee	210
		6.3.2.4	UWB	211

6.3.2.5   LORA 211
6.4   Controller Hardware 212
    6.4.1   The ESP8266/ESP32 212
    6.4.2   Arduino 213
    6.4.3   Raspberry Pi 213
    6.4.4   BeagleBone 214
    6.4.5   Libelium Waspmote 214
    6.4.6   OpenMote 215
6.5   IoT Operating Systems 215
    6.5.1   TinyOS 216
    6.5.2   Contiki/Contiki-NG 217
    6.5.3   FreeRTOS 217
    6.5.4   Android Things 218
    6.5.5   OpenWrt 218
    6.5.6   Raspbian 219
    6.5.7   OpenWSN 219
6.6   IoT Broker 220
    6.6.1   MQTT Brokers 222
        6.6.1.1   Mosquitto 222
        6.6.1.2   RabbitMQ 222
        6.6.1.3   EMQ 222
        6.6.1.4   VerneMQ 223
6.7   Conclusions 223

6.1 OVERVIEW

As proposed in the existing literature, a HyperSurface consists of an array of conductive surfaces interconnected with switching elements as shown in Fig. 6.1. The control HyperSurface's electromagnetic is achieved through the variation of the conductivity of its switching elements [132]. Of course to achieve this, the integration of a network of controllers/gateways is needed. Each controller interacts locally with the HyperSurface with a set of switching elements in order and a set of such controllers are required to control the whole surface. Of course a communication network to connect all these controllers to a central computer is also needed. In other words, in this chapter we assume a 1:N relation between an IoT controller/gateway and any dynamic RF elements in a HyperSurface unit, direct interconnection among them, and direct communication of the controller with the Internet. In this

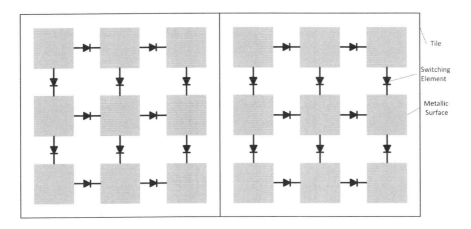

Figure 6.1 Basic structure of a HyperSurface.

section we will provide insights for such a scheme from the IoT aspect, that is the relevant, existing IoT technologies that are available in the market.

6.2 HARDWARE ACTUATION APPROACHES

6.2.1 RF Switching Elements

6.2.1.1 PIN Diodes

PIN diodes are used as current-controlled resistors at RF and microwave frequencies, with resistances that can range from a fraction of an ohm when forward biased or on, to greater than 10 kΩ when reverse biased or off. Differing from typical PN junction diodes, PIN diodes have an additional layer of highly resistive intrinsic semiconductor material (the I in PIN) sandwiched between the P and N material [25]. The structure of a PIN diode is shown in Fig. 6.2.

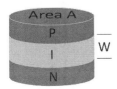

Figure 6.2 Ideal characteristic for an ideal PIN diode.

PIN diodes are mainly used as RF switching elements since their resistance can be controlled by their biasing (supplied current). Their RF resistance is reversely proportional to the supplied current, that is the more current that you inject through the I region, the lower the RF resistance. The current/resistance characteristic is ideally R=K/I (where K is a constant), which looks like a straight line when graphed on log-log scales. A typical PIN R/I graph is shown in Fig. 6.3 [9].

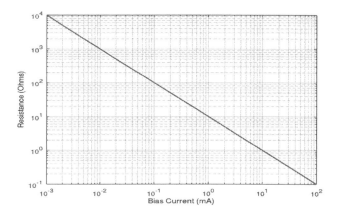

Figure 6.3 Ideal characteristic for an ideal PIN diode.

For a programmable metasurface, the control voltage (bias) for a great number of PIN elements is required. This can be done in an analog or a digital manner. For the analog case, the value of control voltage can be set to any value in a preset rate of values and thus the resistance of the switching element can have a range of values. On the other hand, in the digital case the switching element can be either switched on (nearly zero resistance) or off (open circuit). In both cases a driver circuit is required to supply the biasing current and this circuit should be controlled by the controller.

6.2.2   Controller to PIN Interface

Controller to PIN interface is the system that is used to bias the PIN diode in order to variate its resistance. This can be done either in an analog or in a digital fashion. In both cases a driver circuit is required in order to provide the required current to the diode. This current is applied by a bias driver circuit through inductors that act as an

RF block as shown in Fig. 6.4. The RF signal is passing through the switching element via capacitors acting as a DC block. The control signal generated by the controller is fed in the input of the controller. In the case that analog control over PIN resistance is selected, a Digital to Analog Conversion (DAC) module is required to convert the control value to an analog signal feeding the driver.

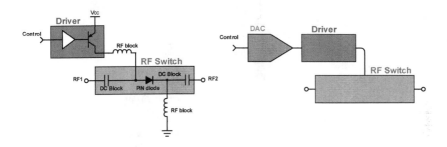

Figure 6.4 PIN basic Biasing circuit (left), Inserting a DAC for analog control (right).

6.2.2.1 DAC

In electronics, a digital to analog converter is a system that converts a digital signal into an analog signal. Due to the complexity and the need for precisely matched components, all but the most specialized DACs are implemented as integrated circuits. There are two basic approaches to implement a DAC module:

■ **Resistor Sum:** This approach is based in creating an adder circuit that adds several input (digital) signals with weights of the power of 2 in the digital inputs. Figure 6.5 (up) shows a 4 input adder in which the gains of the input are designed with a 2^n manner. To achieve that, the resistors of the inputs of the adder should have a value of the power of 2 which makes it difficult to implement with accuracy. One famous topology overcoming this difficulty is the R-2R ladder [11] as shown in Fig. 6.5 (bottom). In this approach each input d_n is added to the output with a weight factor (gain) twice the gain of the previous d_{n-1} input. The R2R ladder uses only 2 values of resistor (R and 2R).

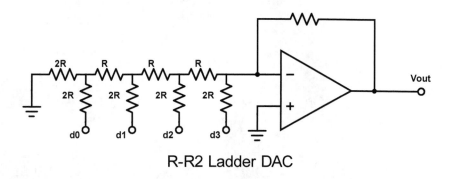

Figure 6.5 Adder based DAC.

■ **PWM:** In this approach the digital value is converted to a Pulse Width Modulated (PWM) signal. A PWM signal is a digital periodic pulse with fixed frequency and a duty cycle proportional to the digital value. This signal is fed into an integrator circuit (e.g., in the simplest case an RC circuit). The output of the integrator is proportional to the duty cycle of the PWM as shown in Fig. 6.5 (bottom). It is worth noting that the common micro-controllers used today for embedded systems (AVR, ESP32 etc.) have hardware implementation of PWM modulators and thus this method for digital to analog conversion is very common and easy to implement. This is due to the fact that only a digital output pin of the micro-controller is required to implement this converter.

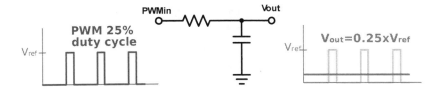

Figure 6.6 DAC based on PWM and integrator.

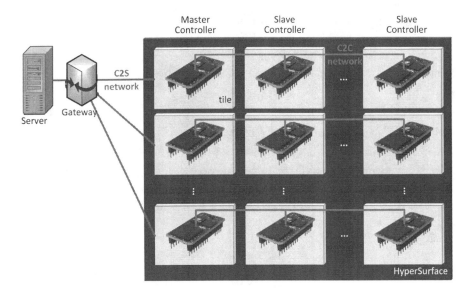

Figure 6.7 Controllers communication architecture.

An important parameter of a DAC system is its resolution that is the number of bits that the DAC takes as input and therefore affects the number of discrete values that the analog voltage can take. As shown in Fig. 6.6, a reference voltage is usually fed in the DAC (or may be internally produced) defines the maximum value of the output. In the case of PWM, the reference voltage is the digital output high voltage.

6.3 CONTROLLER COMMUNICATION

HyperSurfaces are organized in tiles (see Fig. 6.1) where each tile hosts several metallic surfaces. Having this in mind, it is convenient to have one controller placed in each tile responsible to control all the switching elements placed on the tile. In this approach there is the need for a communication network between tiles (controllers) and for a communication network between some controllers and the server from where the Electromagnetic properties of the HyperSurface are controlled. This communication architecture is shown in figure 6.7. In this approach there is one controller per tile. There are master controllers that are connected through a wired communication technology/protocol (see Section 6.3.1) with several slave controllers. This communication channel is shown as a C2C network in Fig. 6.7. The master controller acts as an aggregator for all its slave controllers. Beside controlling its tile switching elements, each master controller is also responsible to communicate the slaves' data to and from the central server that is responsible for controlling the HyperSurface electromagnetic behavior. The technologies that can be utilized for this purpose are explored in Section 6.3.2.

6.3.1   Controller to Controller Communication

Since the distance between controllers is short and tiles' placement on the HyperSurface is fixed a wired network is preferable for the controller to controller communication and since there several controllers to be connected a bus network topology is required. There are several candidate communication protocols for the C2C communication.

6.3.1.1   SPI

The Serial Peripheral Interface (SPI) is a synchronous serial communication interface specification used for short-distance communication, primarily in embedded systems. SPI is a synchronous, full duplex master-slave-based interface. The data from the master or the slave is synchronized with a separate clock signal generated by the master. Both master and slave can transmit data at the same time. The SPI interface can be either 3-wire or 4-wire. In the 4 wire configuration a Chip Select (CS) signal is used to enable the slave device. SPI supports a multi-slave configuration either by using a dedicated CS signal per slave device or in a daisy chain manner. The design of the SPI protocol supports fast data transmission speeds, full duplex communication, and versatile applications in a variety of embedded systems.

6.3.1.2 I2C

I2C is a serial protocol for a two-wire interface to connect low-speed devices like microcontrollers, EEPROMs, A/D and D/A converters, I/O interfaces and other similar peripherals in embedded systems. I2C is not only used on single boards but also to connect components which are linked via cable. Simplicity and flexibility are key characteristics that make this bus attractive to many applications. I2C requires only 2 bus lines, is synchronous and there are no strict baud rate requirements as the master generates a bus clock. It supports bus topology and each device connected to the bus is software-addressable by a unique address. I2C is a true multi-master bus providing arbitration and collision detection [6].

6.3.1.3 UART

UART stands for Universal Asynchronous Receiver-Transmitter and is one of the simplest and oldest forms of device-to-device digital communication. UART is not a communication protocol like SPI and I2C, but a physical circuit in a micro-controller or a stand-alone IC. UARTs transmit data asynchronously, which means there is no clock signal to synchronize the output of bits from the transmitting UART to the sampling of bits by the receiving

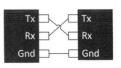

Figure 6.8 UART communication.

UART. Instead of a clock signal, the transmitting UART adds start and stop bits to the data packet being transferred. These bits define the beginning and end of the data packet so the receiving UART knows when to start reading the bits. There are many different speeds supported on UARTs from 300 baud to 115200 baud. UART uses 2 wires and thus it supports full duplex communication as shown in Fig. 6.8.

6.3.1.4 CAN

CAN (also referred to as CANbus or CAN bus) is a network used in many every-day products consisting of multiple micro-controllers that need to communicate with each other. CAN provides a safe communication channel to exchange up to 8 bytes between several network nodes. Additional network functionality like which node talks to which others, when to trigger transmit messages, how to transmit data longer than 8 bytes - all of these functions are specified in so-called higher-layer protocols (in network terms, CAN is a layer 2 implementation - higher layers are implemented in software). CAN supports a bus topology that requires only to cable to operate.

6.3.2 Controller to Server Communication

In the IoT era, several protocols have been designed to support the so-called Machine to Machine (M2M) communication. These communication technologies can be used to connect the master controllers to the server (via the appropriate gateway). In this section the main M2M wireless communication technologies are provided.

6.3.2.1 Bluetooth

The Bluetooth wireless technology was originally conceived in 1994 by Ericsson and first released in 1998 (Bluetooth 1.0). Since then, there have been numerous revisions to the protocol, with version 5 released in 2016. By then, the Bluetooth Special Interest Group had grown to over 30000 members. Bluetooth entails two different operating technologies: BR/EDR (Basic Rate / Enhanced Data Rate), also known as Bluetooth Classic, and BLE (Bluetooth Low Energy). The chosen technology impacts interference and power consumption.

6.3.2.2 802.15.4

802.15.4 is an IEEE standard which targets the physical (PHY) and the medium access (MAC) layers, designed for low-cost / low-complexity devices with low speed requirements. The standard has served as base standard for various other higher layer technologies, including ZigBee, 6LoWPAN and others. 802.15.4 operates in the unlicensed spectrum using three different bands. In practice, the 2.4 GHz band is the most widely used due to its higher speed, which in turn conserves power due to shorted transmission / reception times. The range of 802.15.4 can reach 200 meters during open-air, line-of-sight propagation, and roughly 30 meters in an indoor environment. Table 6.1 lists the 802.15.4 frequencies and available throughput.

Bluetooth operates in the ISM (Industrial Scientific Medical) band, with frequencies between 2.4 and 2.4835 GHz. In order to mitigate interference, Bluetooth utilizes the Frequency Hopping Spread Spectrum (FHSS) technique, where communication occurs on a predetermined sequence of distinct frequencies known to the transmitted and receiver. These distinct frequencies are referred to as channels, and their number varies depending on the operating technology; Bluetooth classic uses 79 channels, while BLE requires 40. The hopping rate is set at 1600 hops per second. In addition to FHSS, Bluetooth employs another technique to combat spectral interference known as Adaptive Frequency Hopping (AFH). In AFH, nodes can detect which channels exhibit high interference and switch to another frequency hopping pattern consisting of less noisy channels. A typical case where AFH comes into play is when one device transmitting at a high bitrate, e.g., a recording video camera, takes up multiple channels for streaming data during a prolonged period of time. Here, other devices sharing the same frequency band will avoid these channels by using AFH to switch to unused channels

Frequency Bands (MHz)	Channels	Throughput (Kbps)	Region
868.3	1	10, 100, 250 (depending on modulation)	Europe
902-928	up to 30 / 2MHz separation	40, 250 (depending on modulation)	North America & Australia
2405-2480	16 / 5MHz separation	250	Worldwide

Table 6.1 802.15.4 Frequencies & Throughput.

Interference avoidance is accomplished using the Carrier Sense Multiple with Collision Avoidance (CSMA/CA) mechanism, which is also used in the 802.11 WiFi standard. In this mechanism, the transmitter listens to a channel and starts transmitting after a predetermined period of time if the channel is idle. The 802.15.4 standard supports two types of devices and two network topologies. The topologies are:

■ Star topology: The simplest form of an 802.15.4 network, requires one node to act as a coordinator, and all other nodes communicate through the coordinating node.

■ Peer-to-Peer topology: In this mesh-type topology, all nodes can communicate with neighbouring nodes directly without the mediation of a coordinating node.

In both topologies, the coordinating node is responsible for setting up the network and sending out beacons when operating in beacon-based mode (see next section). Extensions to the 802.15.4 standard (e.g., Zig-Bee) allow for more complex network topologies (e.g., clusters). Network topologies also dictate the device type of the participating nodes. The possible device types are Full Function Devices (FFD), which can act both as a network coordinator and a simple network participant, and Reduced Function Devices (RFD) which can only participate in a Star topology network, without being able to coordinate it. Within a project context, a Star topology can be used when covering geographically constrained areas with low-cost sensors, whereas if more reliability and larger area coverage is required, a peer-to-peer topology can be considered. However, peer-to-peer to networks are more complex to maintain and can introduce security vulnerabilities.

6.3.2.3 Zigbee

The ZigBee Alliance is an association of companies working to develop and promote a set of open, global standards for a low-power, cost effective wireless personal area network (WPAN) technology based on IEEE 802.15.4. An outcome of this Alliance is ZigBee, a communication protocol which is based on the 802.15.4 specification for its lower PHY and MAC layers. However, the protocol has been designed primarily with mesh networking in mind, therefore it provides additional functionalities in the networking layers in order to provide advanced capabilities such as dynamically forming or self-healing networks.

The allocated frequency bands for ZigBee stem from the 802.15.4 specification and are therefore identical. Slight differences with regards to throughput in the lower frequency bands may exist, given that Zig-Bee is an offshoot of the older 802.15.4 specification released in 2003, and since then the PHY and MAC layers in 802.15.4 have been revised. The main difference to 802.15.4 is how ZigBee treats the networking services on top of these layers. Specifically, ZigBee divides the nodes into three types:

■ Controllers: Similar to the 802.15.4 network coordinator, this type of node is responsible for initiating the network and assigning addresses to participating nodes.

■ Routers: Of interest especially for mesh networking, this type of component handles routing of messages by maintaining routing tables. It can also add and assign addresses and allow new nodes to join the network. Its existence is optional.

■ End Devices: These are the ZigBee equivalent of the 802.15.4 Reduced Functional Devices (RFDs) and can only communicate with controller router nodes. This is also the only node type which can enter sleep mode.

ZigBee uses two types of addresses, long, 64bit addresses unique to the device (and the greater network), and short 16bit addresses which are dynamically assigned by a controller or router node to devices when joining a network. As a consequence, the number of endpoint devices per controller/router nodes is limited to 240, similar to the TCP/IP protocol.

6.3.2.4 UWB

Ultra-WideBand (UWB) is a wireless technology that uses less power and provides higher speed than Wi-Fi and first-generation Bluetooth products. Governed by the WiMedia Alliance, UWB is geared for home theater video, auto safety and navigation, medical imaging and security surveillance. Pulse radio unlike other radio transmissions, UWB does not use a continuous carrier frequency. It transmits extremely short pulses, and the durations between pulses use no power. One method transmits the pulses in continuously varying time slots based on a pseudo-random number sequence like CDMA. The other divides the spectrum into smaller frequency bands that can be added and dropped as necessary. UWB sees through walls because UWB can transmit through materials that would bounce other radio signals; it is also used to pinpoint objects behind barriers or buried underground. First invented by Gerald Ross at Sperry Rand Corporation in the late 1960s, UWB has been used by the military for various radar systems.

6.3.2.5 LORA

LoRa – standing for Long Range - is a Low Powered Wide Area Network technology developed by Semtech. LoRa uses license-free sub-gigahertz radio frequency bands - 868 MHz in Europe - and enables very-long-range transmissions (more than 10 km in rural areas) with low power consumption. In order to deploy a LoRaWAN (i.e., a LoRa-based Wide Area Network), corresponding access points need to be deployed such that coverage is provided over the area of interest. A LoRa network server provides connectivity to other web-based systems (such as cloud-based databases). While the LoRa protocol is a closed, proprietary one, there are commercially available access points and end devices that anyone can procure in order to deploy and manage a

LoRa network. Different users can make use of the same LoRa network server thus providing seamless roaming connectivity across different access points of different owners.

6.4 CONTROLLER HARDWARE

As mention before the HyperSurface is organized in tiles where its tile is controlled by an associated controller. Potentially each controller is an IoT device that connects with its neighboring devices (controller in neighboring tiles) with a wired protocol as described in section. With regards to the implementation of each controller several solutions are mature and available worldwide. The criteria for the hardware selection lies in the size/complexity of the tile, e.g., the number of switching elements and the control type (analog or digital as described in Section 6.2.2). In this section the most popular hardware that is used for IoT applications are given as presented in a survey in [150].

6.4.1 The ESP8266/ESP32

ESP8266 is a well-known Wi-Fi solution among hobbyists and students who want to add an edge of connectivity to their embedded projects (see Fig. 6.9). The ESP8266 is a low-cost Wi-Fi microchips, with a full TCP/IP stack and microcontroller capability, produced by Espressif Systems [5]. It has 1 MB of flash memory, works on the 802.11 b/g/n protocol, and supports Wi-Fi Direct (P2P) and soft-access point. It comes with an

Figure 6.9 ESP 32 module.

integrated transmission-control protocol/Internet protocol stack and a self-calibrated radio-frequency antenna, which allows it to operate under almost all conditions. Several implementations are available in the market based on the ESP8266 chip. The improved version of the ESO8266 is the ESP32. ESP32 is highly-integrated with in-built antenna switches, RF balun, power amplifier, low-noise receive amplifier, filters, and power management modules. ESP32 can perform as a complete standalone system or as a slave device to a host MCU, reducing communication stack overhead on the main application processor. ESP32 can interface with other systems to provide Wi-Fi and Bluetooth functionality through its SPI / SDIO or I2C / UART interfaces.

6.4.2 Arduino

Arduino is an open-source physical computing platform based on a simple i/o board and a development environment that implements the Processing/Wiring language. Arduino offers a vast range of open-source boards capable of performing tasks from blinking an LED to publishing material online to handling heavy networking tasks. This is made possible through the Arduino software integrated development environment (IDE), based

Figure 6.10 Arduino YUN.

on processing. There is a large community of students, hobbyists, and researchers doing projects on Arduino boards and providing many tutorials, and support is available online. For supporting IoT applications, the company offers the Arduino Yun (Fig. 6.10), with onboard Wi-Fi (IEEE 802.11 b/g/n) and Ethernet (IEEE 802.3 10/100Mb/s).

6.4.3 Raspberry Pi

The Raspberry Pi (Fig. 6.11) is a series of small single-board computers developed in the United Kingdom by the Raspberry Pi Foundation. The Raspberry Pi Foundation works to put the power of computing and digital making into the hands of people all over the world. It does this by providing low-cost, high-performance computers that people use to learn, solve problems, and have fun. The original model became far more

Figure 6.11 ESP 32 module.

popular than anticipated, selling outside its target market for uses such as robotics. It does not include peripherals (such as keyboards and mice) or cases. However, some accessories have been included in several official and unofficial bundles. The Raspberry Pi is a very cheap computer that runs Linux, but it also provides a set of GPIO (general purpose input/output) pins that allow you to control electronic components for physical computing and explore the Internet of Things (IoT). The Raspberry Pi operates in the open source ecosystem: it runs Linux (a variety of distributions), and its main supported operating system, Raspbian, is open source and runs a suite of open source software. The latest version is the Raspberry Pi 4 which among others features Gigabit Ethernet, IEEE 802.11ac wireless and Bluetooth 5.0, BLE interfaces while processing power-wise it is built upon a Quad core Cortex-A72 (ARM v8) 64-bit SoC with up to 4 GB RAM.

6.4.4 BeagleBone

The BeagleBoard [1] is a low-power open-source single-board computer pro-duced by Texas Instruments in associ-ation with Digi-Key and Newark ele-ment14. The BeagleBoard was also designed (see Fig. 6.12) with open source software development in mind, and as a way of demonstrating the Texas Instrument's OMAP3530 system-on-a-chip. Several variations of the Bea-gleBoard hardware have been available. The BeagleBoards can run a grow-ing list of operating systems including

Figure 6.12 BeagleBone black.

Linux, FreeBSD, OpenBSD, RISC OS and SymbianOS, with support for Android in active development. Newer models also include onboard Wi-Fi networking. Several peripherals are included in the BeagleBoard hardware such as UARTs, SPI, I2C, CANbus, and of course IO pins as well as optional expansion boards are available. BeagleBone is the latest series of the BeagleBoard devices.

6.4.5 Libelium Waspmote

The Libelium Waspmote is a sensor device platform specially (see Fig. 6.13) oriented to systems integrators and developers. The device family supports a wide range of sensor technologies, expan-sion boards and communications proto-cols. The codebase is extensive, well doc-umented and has proven reliability and stability in the field. Waspmote solu-tions can be customized to your needs, incorporated into your housing or envi-ronment – and the code stack devel-oped with your application in mind. Waspmote will accelerate time to mar-ket and by using a device from Libelium,

Figure 6.13 Libelium Waso-mote.

it means you benefit from their extensive certification, testing, software updates and sensor comparability. The devices support different communication protocols (ZigBee, Bluetooth, 3G/GPRS) and frequencies (2.4GHz, 868MHz, 900MHz) – perfect for a solution targeting a global market where regulations can vary by country. Communication links of up to 12KM can be established using different protocols (i.e., LoRa), or far higher using technologies such as 3G/4G.

6.4.6 OpenMote

OpenMote is a open-hardware proto-typing ecosystem designed to accelerate the development of the Industrial Internet of Things (IIoT). It features the OpenMote-CC2538, IIoT, see Fig. 6.14 a state-of-the-art computing and communication device. This device interfaces with several other accessories or "skins", through a standardized connector. The skins developed to date include boards to provide power, boards which enable a developer to easily debug the platform, and boards to allow seamless integration of an OpenMote network into the Internet. This hardware ecosystem is comple-

Figure 6.14 Libelium Waso-mote.

mented by a suite of software tools and ports to popular open-source IoT implementations. The OpenMote platform is for example tailored to run the OpenWSN open-source implementation of emerging IIoT standards. The combination of hardware and software ecosystems gives an embedded programmer an intuitive and complete development environment, and an end-user a fully working low-power wireless mesh networking solution running the latest IIoT standards [173].

6.5 IOT OPERATING SYSTEMS

For the operation of the controller, especially the master ones, an operating system may be required to run the tasks of the status/control data. Legacy operating systems (OS) like Linux are capable to fulfill the diverse requirements of IoT systems, which include heterogeneous hardware constraints, as well as various network stack, autonomy and real-time constraints. Although several efforts aimed at adapting existing OS for the IoT, key features such as maximum energy efficiency or strong real-time guarantees cannot be efficiently implemented as add-ons to a pre-existing system, because such features impact every part of the system. The diversity of the requirements that need to be fulfilled by software running on IoT devices is too challenging for existing OSs, which were not designed to run on the full range of hardware platforms that compose the IoT and are having storage processing power and energy constraints. Thus several OSs specifically developed to fulfill the IoT devices' needs. In this section we explore the most popular IoT operating systems.

6.5.1   TinyOS

TinyOS [12] is an open source, BSD-licensed operating system designed for low-power wireless devices, such as those used in sensor networks, ubiquitous computing, personal area networks, smart buildings, and smart meters. TinyOS  has been adopted by thousands of developers worldwide, on many platforms, for a broad range of wireless sensor networks. Its communication-centric design and modular software model are tailored to the unique requirements of these networks, where applications and services are distributed over collections of resource-constrained, unattended application-specific devices streaming data to and from the physical world. From its roots in the academic research community, it is emerging as an important vehicle for commercial deployments. TinyOS was designed specifically for WSNs. It introduces a structured event-driven execution model and a component-based software design that supports a high degree of concurrency in a small footprint, enhances robustness, and minimizes power consumption while facilitating implementation of sophisticated protocols and algorithms. The system and its services comprise components connected with well-defined interfaces, much as a schematic wires hardware blocks together. The diversity of hardware platforms, protocols, and applications is addressed by plugging together the necessary components from a catalog of candidates [13]. Communications and networking have driven the TinyOS design as available radios and microcontroller interfaces evolved. The communication subsystem has to meet real-time requirements and respond to asynchronous events while other devices and processes are serviced. Early designs modulated the radio channel bit-by-bit in software, while the Mica and Mica2 generations used a byte-level interface. Most recent platforms use IEEE 802.15.4 radios, such as the ChipCon CC2420, with a packet-level interface, but with differing microcontroller-radio interconnections. The TinyOS 2.0 radio component provides a uniform interface to the full capabilities of the used radio chip. TinyOS has a component-based programming model, codified by the nesC language, a dialect of C. TinyOS is not an OS in the traditional sense; it is a programming framework for embedded systems and set of components that enable building an application-specific OS into each application [88].

6.5.2 Contiki/Contiki-NG

Contiki is an operating system for networked, memory-constrained systems with a focus on low-power wireless Internet of Things devices. Extant uses for Contiki include systems for street lighting, sound monitoring for smart cities, radiation monitoring, and alarms. It is open-source software  released under a BSD license. Contiki-NG is an improved version of the contiki, an operating system for resource-constrained devices in the Internet of Things. Contiki provides multitasking and a built-in Internet Protocol Suite (TCP/IP stack), yet needs only about 10 kilobytes of random-access memory (RAM) and 30 kilobytes of read-only memory (ROM). A full system, including a graphical user interface, needs about 30 kilobytes of RAM.

A new branch has recently been created, known as Contiki-NG. Contiki-NG contains an RFC-compliant, low-power IPv6 communication stack, enabling Internet connectivity. The system runs on a variety of platforms based on energy-efficient architectures such as the ARM Cortex-M3/M4 and the Texas Instruments MSP430. The code footprint is on the order of a 100 kB, and the memory usage can be configured to be as low as 10 kB. The source code is available as open source with a 3-clause BSD license.

6.5.3 FreeRTOS

FreeRTOS is a real-time operating system kernel for embedded devices. FreeRTOS is designed to be small and simple. The kernel itself consists of only  three C files. To make the code readable, easy to port, and maintainable, it is written mostly in C, but there are a few assembly functions included where needed. It is an open source real-time operating system (for embedded systems). FreeRTOS supports many different architectures and compiler toolchains, and is designed to be "small, simple, and easy to use". Like all operating systems, FreeRTOS's main job is to run tasks. Most of FreeRTOS's code involves prioritizing, scheduling, and running user-defined tasks. Unlike all operating systems, FreeRTOS is a real-time operating system which runs on embedded systems. FreeRTOS ships with all the hardware-independent as well as hardware-dependent code you'll need to get a system up and running. It supports many compilers (CodeWarrior, GCC, IAR, etc.) as well as many processor architectures (ARM7, ARM Cortex-M3, various PICs, Silicon Labs 8051, x86, etc.). FreeRTOS implements multiple threads by having the host program call a thread tick method at regular short intervals. The thread tick method switches tasks depending on priority and a round-robin scheduling scheme.

6.5.4 Android Things

Android Things (codenamed Brillo) is an Android-based embedded operating system platform by Google, announced at Google I/O 2015. It is aimed to be used with low-power and memory constrained Internet of Things (IoT) devices, which are usually built from different MCU platforms. As an IoT OS  it is designed to work as low as 32–64 MB of RAM. It will support Bluetooth Low Energy and Wi-Fi. Along with Brillo, Google also introduced the Weave protocol, which these devices can use to communicate with other compatible devices. Android uses the Linux kernel at its core and Linux is a full multi-tasking operating system with virtual memory support. This means that Android Things needs a processor that supports virtual memory, in other words a processor with a full MMU.

6.5.5 OpenWrt

The OpenWrt Project [8] is a Linux operating system targeting embedded devices. Instead of trying to create a single, static firmware, OpenWrt provides a fully writable filesystem with package man-  agement. This frees you from the application selection and configuration provided by the vendor and allows you to customize the device through the use of packages to suit any application. For developers, OpenWrt is the framework to build an application without having to build a complete firmware around it; for users this means the ability for full customization, to use the device in ways never envisioned.

OpenWrt can be ported to a variety of hardware devices but mainly it targets to operate on access point devices which makes it suitable, for the IoT use-case, to build smart gateways that can support a variety of wireless access technologies.

6.5.6 Raspbian

As its name dictates, Raspbian is a free operating system based on Debian, optimized for the Raspberry Pi hardware. There are several versions of Raspbian including Raspbian Buster and Raspbian Stretch. Since 2015 it has been officially provided by the Raspberry Pi Foundation as the primary operating system for the family of Raspberry

Pi single-board computers. Raspbian uses PIXEL, Pi Improved X-Window Environment, Lightweight as its main desktop environment as of the latest update. It is composed of a modified LXDE desktop environment and the Openbox stacking window manager with a new theme and few other changes.

6.5.7 OpenWSN

OpenWSN is a project created at the University of California Berkeley and extended at the INRIA and at the Open University of Catalonia (UOC) which aims to build an open standards-based and open source implementation of a complete constrained network protocol stack for wireless sensor networks and Internet of Things. The root of OpenWSN is

a deterministic MAC layer implementing the IEEE 802.15.4e TSCH based on the concept of Time Slotted Channel Hopping (TSCH). Above the MAC layer, the Low Power Lossy Network stack is based on IETF standards including the IETF 6TiSCH management and adaptation layer (a minimal configuration profile, 6top protocol and different scheduling functions). The stack is complemented by an implementation of 6LoWPAN, RPL in non-storing mode, UDP and CoAP, enabling access to devices running the stack from the native IPv6 through open standards.

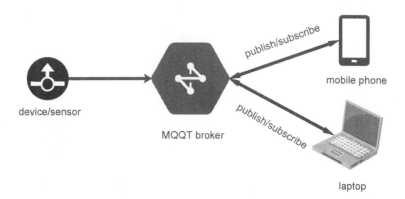

device/sensor

publish/subscribe

mobile phone

MQQT broker

publish/subscribe

laptop

Figure 6.15 Controllers communication architecture.

6.6 IOT BROKER

In the world of the IoT several sensors and actuators are collecting data that need to be exchanged in an efficient and resilient manner. IoT devices generate messages, like status reports and environmental measurements and they receive actuation commands. Moreover, the number of IoT devices is growing incredibly fast and only their combined message traffic is growing faster. And as well as the explosion of the IoT, there are applications running on smartphones, tablets, and computers that all need the same type of messaging service.

Probably the most widely adopted standard in the Industrial IoT to date, Message Queuing Telemetry Transport (MQTT) is a lightweight publication/subscription type (pub/sub) messaging protocol. Designed for battery-powered devices, MQTT's architecture is simple and lightweight, enabling low power consumption for the connected devices. Working on top of the TCP/IP protocol, it has been especially designed for unreliable communication networks in order to respond to the problem of the growing number of small-sized cheap low-power objects that often appear in IoT networks. MQTT is based on a **subscriber**, **publisher**, model as presented in Fig. 6.15. Within the model, the publisher's task is to collect the data and send information to subscribers via the mediation layer (broker). The role of the broker, on the other hand, is to ensure security by cross-checking the authorization of publishers and subscribers. In MQTT are three levels of Quality of Service (QoS):

■ QoS0 (At most once): The least reliable mode but also the fastest. The publication is sent but confirmation is not received.

■ QoS1 (At least once): Ensures that the message is delivered at least once, but duplicates may be received.

■ QoS2 (Exactly once): The most reliable mode while the most bandwidth-consuming. Duplicates are controlled to ensure that the message is delivered only once.

The client that publishes the message to the broker defines the QoS level of the message when it sends the message to the broker. The broker transmits this message to subscribing clients using the QoS level that each subscribing client defines during the subscription process. Having found wide application in such IoT devices as electric meters, vehicles, detectors, and industrial or sanitary equipment, MQTT responds well to the following needs:

■ Minimum bandwidth use

■ Operation over wireless networks

■ Low energy consumption

■ Good reliability if necessary

■ Little processing and memory resources

A variety of MQTT is MQTT-SN (MQTT for Sensor Networks). MQTT-SN is aimed at embedded devices on non-TCP/IP networks, whereas MQTT itself explicitly expects a TCP/IP stack. Despite its characteristics, MQTT can be problematic for some very restrictive devices, due to the fact of the transmission of messages over TCP and managing long topic names. This is solved with the MQTT-SN variant that uses UDP and supports topic name indexing. However, despite its wide adoption, MQTT doesn't support a well-defined data representation and device management structure model, which hinders the implementation of its data management and device management capabilities.

6.6.1 MQTT Brokers

6.6.1.1 Mosquitto

Eclipse Mosquitto [4] is an open source (EPL/EDL licensed) message broker that implements the MQTT protocol versions 5.0, 3.1.1 and 3.1. Mosquitto is lightweight and is suitable for use on all devices from low power single board computers to full servers. The Mosquitto project also provides a C library for implementing MQTT clients, and the very popular mosquittopub and mosquittosub command line MQTT clients. Mosquitto is part of the Eclipse Foundation, is an iot.eclipse.org project and is sponsored by cedalo.com. Mosquitto is a really lightweight MQTT broker supporting TLS encryption and there are plugins for authorization using a database.

6.6.1.2 RabbitMQ

RabbitMQ [10] is an open-source message-broker software that originally implemented the Advanced Message Queuing Protocol (AMQP) and has since been extended with a plug-in architecture to support Streaming Text Oriented Messaging Protocol (STOMP), MQ Telemetry Transport (MQTT), and other protocols. The RabbitMQ server program is written in the Erlang programming language and is built on the Open Telecom Platform framework for clustering and failover. Client libraries to interface with the broker are available for all major programming languages. RabbitMQ has TLS and clustering support.

6.6.1.3 EMQ

EMQ [7] is another Erlang based broker which was very promising. EMQ X Broker is a distributed, massively scalable, highly extensible MQTT message broker written in Erlang/OTP. EMQ X Broker is previously known as emqtt, since R3.1, the name is officially changed to EMQ X Broker. It supports MQTT V3.1/V3.1.1 and V5.0 protocol standards. With the plugins, it can support MQTT-SN, WebSocket, CoAP, LwM2M, Stomp and other proprietary protocols based on TCP/UDP. EMQ X works as single broker node or cluster.

It provides scalable, reliable, MQTT message interconnection for IoT, IoV, M2M, Smart Hardware and Mobile Messaging Applications.

6.6.1.4 *VerneMQ*

VerneMQ [2] is a relatively new MQTT high-performance, distributed MQTT broker. It scales horizontally and vertically on commodity hardware to support a high number of concurrent publishers and consumers while maintaining low latency and fault tolerance. VerneMQ is the reliable message hub for your IoT platform or smart products. VerneMQ is designed from the ground up to  work as a distributed message broker, ensuring continued operation in the event of node or network failures and easy horizontal scalability. VerneMQ uses a master-less clustering technology. There are no special nodes like masters or slaves to consider when the inevitable infrastructure changes or maintenance windows require adding or removing nodes. This makes operating the cluster safe and simple.

6.7 CONCLUSIONS

A HyperSurface consists of several metallic surfaces interconnected via high frequency switching elements. To variate the electromagnetic properties of the HyperSurface, these switching elements have to change state, i.e., open, close, or have a specific impedance. To control the switch state, a network of controllers is required and these controllers have to be interconnected to accept commands by a central server. In this chapter, the implementation of the HyperSurface physical control was explored. An introduction on the high frequency switching elements was given and the control of such elements was outlined. Then the communication between controllers and server was examined. Next, an overview of the most popular solutions with regards to controller hardware among the available operating systems was given. Finally, a presentation of the higher level (application) messaging protocol that is commonly used in IoT applications was provided with the most popular implementation of it.

CHAPTER 7

Interim: Drafting a Stack

Christos Liaskos

Foundation for Research and Technology Hellas, 71110, Heraklion, Crete.
Email: cliaskos@ics.forth.gr - corresponding author

In the preceding chapters, the authors presented the various components of a programmable metasurface. At this point it worth noting a resemblance that begins to appear in terms of organization in the form of a components and functionalities stack.

First, in Chapter 3 we saw the electromagnetic aspect of such a material. In other words, the authors showed the physical principles that such a material should adhere to, in order to be able to manipulate waves. Second, in Chapter 4 we saw the software modeling aspect, that is the data objects and workflows that can be used to describe the physical functionalities, and how they can be invoked using a set of software classes. Third, in Chapter 5 the authors showed the networking protocols that can be used for communicating with a programmable material over a network. Fourth, in Chapter 6 the author presented existing IoT approaches for implementing this communication, and software mechanisms.

As shown in Fig. 7.1, one could attempt to organize these functionalities in the form of a draft stack, resembling, e.g., the OSI stack. At the bottom, we could place all existing IoT platforms that can be used for implementing the described control approaches. In correspondence to the OSI stack, this, along with the electromagnetic aspect could comprise the physical layer of the studied materials. Since IoT systems provide a wide array of networking solutions, the network part could potentially be implemented within this layer. However, from a logical point of view, the network functionality comes on top of the physical

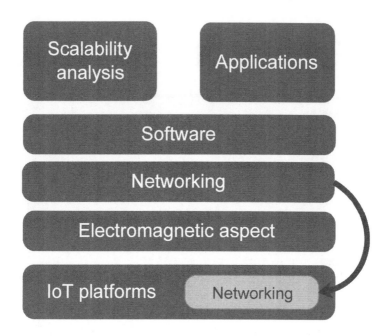

Figure 7.1 A draft organization of programmable metamaterial components in a form of a stack.

layer. The software layer is then placed over the facilities offered by the preceding layers.

Finally, in the ensuing chapters we will move on to study the scalability laws and applications of the programmable materials.

This organization is of course a draft and does not follow a standard. However, it is provided to help in exposing, describing and digesting the preceding and ensuing material.

CHAPTER 8

The Scaling Laws of HyperSurfaces

Hamidreza Taghvaee, Sergi Abadal, Eduard Alarcon, and Albert Cabellos-Aparicio

NaNoNetworking Center in Catalonia (N3Cat), Universitat Politecnica de Catalunya, 08034 Barcelona, Spain

Taqwa Saeed and Andreas Pitsillides

Department of Electrical and Computer Engineering, University of Cyprus, P.O. Box 20537, 1678, Nicosia, Cyprus

Odysseas Tsilipakos, Christos Liaskos, Anna Tasolamprou, Maria Kafesaki, Alex Pitilakis and Nikolaos Kantartzis

Foundation for Research and Technology Hellas, 71110, Heraklion, Crete, Greece

Vassos Soteriou

Department of Electrical Engineering and Computer Engineering and Informatics, Cyprus University of Technology, Limassol, Cyprus

Marios Lestas

Electrical Engineering Department, Frederick University, Nicosia, Cyprus

CONTENTS

8.1   The HyperSurface Scalability versus Manufacturing
      Technologies 229
      8.1.1   Scaling Model 231
              8.1.1.1   Dimensional Factors 231
              8.1.1.2   Programming Parameters 232

8.1.2   Methodology 232
     8.1.2.1   Unit Cell Model 234
     8.1.2.2   Metasurface Model 235
     8.1.2.3   Metasurface Coding 237
     8.1.2.4   Performance Metrics 238
     8.1.2.5   Validation 239
8.1.3   Performance Scalability 241
     8.1.3.1   Directivity 241
     8.1.3.2   Target Deviation 242
     8.1.3.3   Half Power Beam Width 243
     8.1.3.4   Side Lobe Level 244
8.1.4   The HyperSurface Energy Footprint, Cost, and
     Performance 244
     8.1.4.1   Cost and Power Models 245
     8.1.4.2   Application-Specific Figures of Merit . 248
     8.1.4.3   Performance-Cost Analysis 249
8.2   The HyperSurface Data Traffic as a Scaling Concern 249
  8.2.1   System Model 252
     8.2.1.1   Mobility Model 253
     8.2.1.2   Gateway Model 254
     8.2.1.3   Embedded Controller Network 254
  8.2.2   Evaluation Methodology 255
     8.2.2.1   Relevant Inputs 256
     8.2.2.2   Traffic Analysis Metrics 256
     8.2.2.3   Walkthrough Example 257
  8.2.3   Workload Characterization 257
     8.2.3.1   Spatio-Temporal Intensity 257
     8.2.3.2   Reconfiguration Delay 260
     8.2.3.3   Sensitivity Analysis 261
  8.2.4   Indoor Mobility Scenario 263
8.3   Conclusions 265

HyperSurfaces integrate a network of controllers within the structure of the metasurface (MS) [130]. Controllers drive the reconfigurable unit cells and exchange information with neighbouring controllers so that the HyperSurface (HSF) can i) implement a given electromagnetic functionality requested by an authorized user, and ii) adapt to changes in the environment. The internal network of controllers is the enabler of the HSF approach and the main difference with respect to a conventional programmable MS.

One of the many challenges posed by this novel approach concerns the design and development of the controllers and the interconnection network within the HSF [85, 140, 162]. Such a task is largely hindered by the unique combination of resource constraints and communication requirements of this new networking scenario, which prevents the use of conventional techniques and requires radically new solutions instead. Moreover, the network design may need to adapt to the different HSF use cases and evolve as soon as technology advances enable the creation of HSF with higher performance and capable of operating up to the THz or optical regime [15, 196].

It is very difficult to design the HSF network (or any network in general) if the traffic is not quantified and the relevant performance metrics are not clarified. This chapter aims to bridge the aforementioned gap. We essentially focus on the relationship between application requirements and MS design. Such a link is important to adequately dimension the intra-HSF network and, more importantly, characterize the traffic that such network will have to support. From this information, we can derive guidelines to drive the whole design process of future HSFs.

To facilitate the design of programmable metasurfaces, we study them from two different perspectives: first, Section 8.1 reviews the general architecture including energy, cost and performance of the HSF paradigm with scalability analysis. Then, Section 8.2 depicts the traffic analysis performed to better handle the design of the network that is integrated within the structure of the HSF. The chapter is concluded in Section 8.3.

8.1 THE HYPERSURFACE SCALABILITY VERSUS MANUFACTURING TECHNOLOGIES

The design of a programmable MS faces significant challenges in terms of complexity due to the many aspects involved, including but not limited to the unit cell design, the placement of the tuning elements and its impact on the unit cell response, the tuning range requirements, the integration of driving methods within the MS structure or the associated extra fabrication steps [107]. Furthermore, mapping all these design aspects to the requirements of the functionalities to be implemented by the programmable MS is an arduous task.

In the pathway to dimensioning a programmable MS for a given functionality, a key question is then: **how do the metasurfaces scale?** In more detail, we should try to know how do metasurfaces scale in terms of performance for the target functionality. Thus, a scalability analysis would inspect the impact of each design parameter to the metasurface performance.

This section revolves around the concept of scalability analysis of HSFs, in an attempt to provide tools to derive general considerations and guidelines of design for future metasurfaces. For instance, a clear outcome of this analysis would be the answer to the question:

■ **What is the minimum number of unit cells, number of states per unit cell, and maximum unit cell size that guarantee a given performance for a given EM functionality?**

In the pathway to answering this question, we will face several tradeoffs that we can navigate easily thanks to the analytical modeling of the metasurface. For instance, we shall be able to answer the (also pertinent) question, among many others:

■ **Is it preferable to improve in terms of discretization of the space (with smaller unit cells) or in terms of quantization of the phases (with more states) to comply with the requirements of a given application?**

Coupled to complexity or cost models or tied to other tools such as information theory methods [44,183], the scalability analysis could clarify the practicable space of HSF design, illustrating the main design tradeoffs and delimiting optimal design regions. In particular, the section develops the methodology for scalability analysis and applies it for a representative functionality like anomalous reflection for beam steering.

The rest of the section is structured as follows. The scaling model is described in Section 8.1.1, whereas the main methodology aspects are detailed in Section 8.1.2. The results of the scalability analysis are given in Section 8.1.3 and a discussion of the contradicting trends between performance and cost is delivered in Section 8.1.4.3.

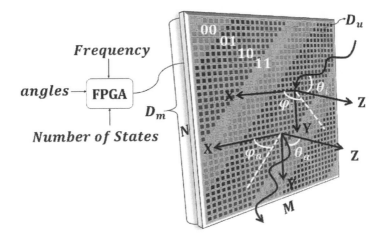

Figure 8.1 Schematic representation of a programmable metasurface implementing a phase gradient for beam steering applications.

8.1.1 Scaling Model

Figure 8.1 shows a schematic representation of the HSF under study. We assume that the HSF has $N \times M$ reconfigurable unit cells and is driven by a control sub-system whose function is to change the unit cell states to point the reflected wavefront to a given direction. In the scenario explored in this study, the control sub-system takes as inputs the incidence (θ_i, ϕ_i) and target reflection angles (θ_r, ϕ_r), as well as the number of states N_s that the unit cells can adopt. Due to the limited number of states, the HSF may reflect the wave towards a direction (θ_a, ϕ_a) that is a bit off-target. The next sections describe the main parameters in a bit more detail.

8.1.1.1 Dimensional Factors

- **Size of the unit cell (D_u):** The unit cell dimensions generally depend on the desired frequency band. Assuming, as we will see, that the unit cell design provides the necessary amplitude or phase, we emphasize the impact of the unit cell dimensions on the performance of the HSF. We assume that the unit cell shape is square of side D_u.

- **Size of the metasurface (D_m):** The application normally defines certain size requirements to, for instance, cover an entire surface. However, the HSF area may be limited by power consumption or cost. Here, we assume that $M = N$ and, therefore, that the MS covers a square area with lateral size of D_m.

■ **Wavelength (λ):** From the electromagnetic point of view, determining the frequency band of interest is critical to tackle the design of the unit cell. However, from the scalability perspective and again assuming that the unit cells respond as expected, the importance lies in the size of the unit cell and the size of the metasurface normalized to the wavelength, i.e., unit cells need to be subwavelength and the metasurface should cover several wavelengths to be effective.

8.1.1.2 Programming Parameters

■ **Number of unit cell states (N_s):** Ideally, an HSF would attain continuous control over the amplitude and phase of the unit cell responses. However, complexity issues related to the tuning elements and their driving methods lead to a discretization of the unit cell response. Also, as we will see, discretization may have a very limited impact on the performance of the metasurface. In the study, a parameter N_s models the number of unit cell states offered by the control sub-system.

■ **Target direction (θ_r, ϕ_r):** The direction of reflection stands as a clear application-specific requirement for beam steering. In fact, the unit cell states necessary for obtaining the desired reflection direction depend on both incidence and reflected directions. Therefore, an HSF targeting beam steering shall have the incidence and reflection directions as programming inputs. However, without loss of generality, we assume normal incidence ($\theta_i = \phi_i = 0$) to keep the analysis tractable and leave the target direction of reflection as the main parameter. To this end, we use the angles (θ_r, ϕ_r) as we can express the position of the emitter and intended receiver with spherical coordinates (r, θ, ϕ) using the MS as point of reference in the coordinate system as shown in Fig. 8.1.

8.1.2 Methodology

Given an HSF with an arbitrary number of unit cells, the Floquet periodic boundary condition models an infinite periodicity of the unit cell along the direction normal to the specified boundaries. In general, this

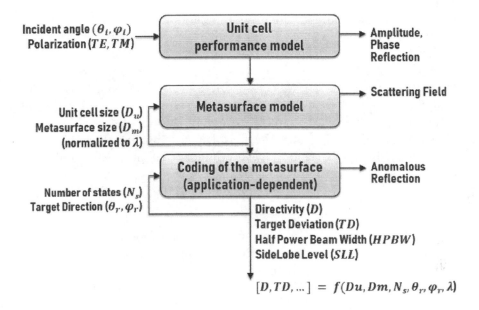

Figure 8.2 Flowchart of the proposed methodology instantiated for beam steering. Unit cell models (Section 8.1.2.1) provide amplitude and phase of individual cells, to be plugged into a metasurface model (Section 8.1.2.2) that calculates the scattering field of a particular metasurface design. The metasurface coding (Section 8.1.2.3) process gives the unit cell states necessary to achieve a given functionality described by a set of performance metrics (Section 8.1.2.4).

simplifies the structure model and reduces the computational burden related to the full-wave electromagnetic simulation of the HSF. However, such method only applies to static designs or functionalities that only need global tuning, i.e., all unit cells biased to the same set of values, as the Floquet periodic boundary does not allow to break the periodicity.

In the case of functionalities requiring local tuning such as anomalous reflection, unit cells can have a state that differs from that of their neighbors. Therefore, the periodic boundary condition is not valid anymore and we have to model the entire structure. However, solving Maxwell's equations in a relatively large metasurface with small unit cells becomes extremely intensive in terms of memory and computation time. This is clearly not compatible with a scalability analysis which involves hundreds of simulations for different combinations of the input parameter values. As such, we need a methodology that is fast and reasonably accurate to expose the scalability trends of metasurfaces. Here, we use analytical models to that end.

Figure 8.2 summarizes the proposed methodology and exemplifies it for the case of anomalous reflection. Sections 8.1.2.1 and 8.1.2.2 depict the unit cell and metasurface models, respectively. Sections 8.1.2.3 and 8.1.2.4 outline the methods used to derive the optimal coding of the HSF for beam steering and the performance metrics that this work considers. Finally, Section 8.1.2.5 validates the proposed analytical approach and MS coding method.

8.1.2.1   Unit Cell Model

Unit cells are generally designed with a certain function in mind. For instance, the anomalous reflection will require unit cells to exhibit a reflection coefficient with high amplitude plus reconfigurable phase to change the angle of reflection [193]. As shown in several works, providing phase reconfigurability can be achieved via several tuning mechanisms [109]. Here we provide a particular unit cell design.

We assume a square unit cell ($c = 2$ mm) with a metallic backplane for operation around 25 GHz, aimed at giving service to one of the available $5G$ bands according to new recommendations by the International Telecommunication Union (ITU) [171]. A square metallic patch ($b = 1.85$ mm) is stacked on top of a substrate (Rogers RO4003C) with permittivity $\epsilon_r = 3.5$ and thickness $a = 0.81$ mm. It is possible to modify the phase response of the unit cell by adding capacitance to the square metallic patch. For phase tunability, this capacitance is given by varactors, which are embedded within the controllers and hidden under the backplane, but connected to the top patch with vertical vias [162].

Unit cell design at 25 GHz has some challenges, including small dimensions and the need for utilizing very sensitive yet extremely small tuning elements (e.g., varactors). In comparison to previous designs [109], our proposed unit cell design (Figure 8.3) uses multiple varactors per unit cell to reduce the sensitivity. More importantly, instead of incorporating a floating element at the center, it is preferred to put four elements grounded with die chips ($d = 0.2mm$, $e = 1.37mm$) to avoid floating elements which can cause serious problems at the electronics level.

We implement the proposed unit cell in a full-wave solver, CST MWS [3], and evaluate the reflection coefficient when the unit cell is illuminated by a normal incident plane wave and for a set of capacitance values. Assuming that our design implements four coding states,

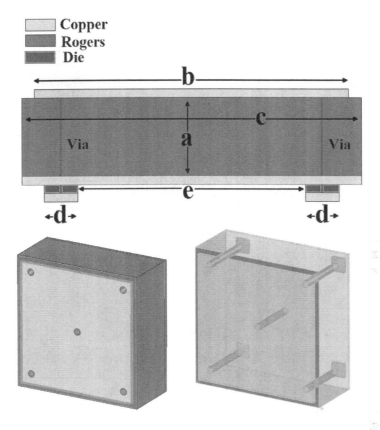

Figure 8.3 Cross-section, top-view, and bottom-view of the assumed unit cell.

it is standard practice in anomalous reflection MS that the 2π phase range is divided into evenly spaced states with $\pi/2$ separation with high reflection amplitude [109]. As shown in Figure 8.4, the unit cell achieves these objectives around the target frequency, 25 GHz, with a reflection amplitude Γ of 0.9 and phases Φ at $\{45, 135, 225, 315\}$ degrees. The figure plots the capacitances that have achieved such separation: 0.01 pF, 0.04 pF, 0.06 pF, and 0.9 pF. We will see that, if the capacitances deviate from such values, the unit cell may inaccurately point to different amplitude and phase.

8.1.2.2 Metasurface Model

Beam steering is a particular case of wavefront manipulation that occurs in the far field. As such, the metasurface can be accurately

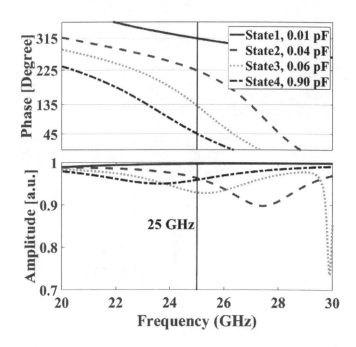

Figure 8.4 Unit cell reflection phase Φ (top) and amplitude Γ (bottom) as a function of frequency for the four chosen capacitance values.

modeled as a compact array following the Huygens principle [22]. This method has been validated in several works via extensive simulations [64]. Considering each unit cell as an element of the array, the far field of the metasurface can be obtained as

$$F(\theta, \phi) = f_E(\theta, \phi) \cdot f_A(\theta, \phi), \tag{8.1}$$

where θ is the elevation angle, ϕ is the azimuth angle of an arbitrary direction, $f_E(\theta, \phi)$ is the element factor (pattern function of unit cell) and $f_A(\theta, \phi)$ is the array factor (pattern function of unit cell arrangement). With the widespread assumption of a planar wave covering the entire metasurface, the scattering pattern will depend only on the array factor. For the metasurface with $N \times M$ unit cells, the far field pattern becomes

$$F(\theta, \phi) = \sum_{m=1}^{M} \sum_{n=1}^{N} A_{mn} e^{j\alpha_{mn}} f_{mn}(\theta_i, \phi_i)$$
$$\Gamma_{mn} e^{j\Phi_{mn}} f_{mn}(\theta, \phi) e^{jk_0 \zeta_{mn}(\theta, \phi)} \tag{8.2}$$

where A_{mn} and α_{mn} are the amplitude and phase of the wave incident to the mn-th unit cell; Γ_{mn} and Φ_{mn} are the amplitude and phase of the response of the mn-th unit cell; $f_{mn}(\theta, \phi)$ denotes the scattering diagram of the mn-th unit cell towards an arbitrary direction of reflection, whereas $f_{mn}(\theta_i, \phi_i)$ denotes the response of the mn-th unit cell at the direction of incidence determined by θ_i, ϕ_i and $k_0 = 2\pi/\lambda_0$ is the wave number (air is assumed). Finally, $\zeta_{mn}(\theta, \phi)$ is the relative phase shift of the unit cells with respect to the radiation pattern coordinates, given by

$$\zeta_{mn}(\theta, \phi) = D_u \sin\theta[(m - \tfrac{1}{2})\cos\phi + (n - \tfrac{1}{2})\sin\phi]. \tag{8.3}$$

We further make the plausible assumption of plane wave incidence, so that factors A_{mn}, α_{mn}, and $f_{mn}(\theta_i, \phi_i)$ are constants for all m, n.

Further, we model the scattering pattern of the unit cell over the positive semisphere with the function $\cos(\theta)$, which is a widespread assumption [188]. Finally, and without loss of generality, we consider the normal incidence ($\theta_i = \phi_i = 0$). Then, Eq. (8.2) becomes

$$E(\theta, \phi) = K\cos\theta \sum_{m=1}^{M}\sum_{n=1}^{N}\Gamma_{mn}e^{j[\Phi_{mn}+k_0\zeta_{mn}(\theta,\phi)]} \tag{8.4}$$

where K is a constant. By controlling the phase shift of the unit cells Φ_{mn}, we can implement anomalous reflection as described next.

8.1.2.3 Metasurface Coding

The coding of the cells of a programmable metasurface allows obtaining the desired functionality. In other words, we need to derive the amplitude Γ_{mn} and phase Φ_{mn} of each unit cell so that the collective response matches with the required functionality. Then, we map the required Γ and Φ to the closest available unit cell states.

In the case of anomalous reflection for beam steering, analytical methods provide high accuracy. In this work, we follow the well-known principles of wavefront manipulation whereby a phase gradient is used to determine the direction of reflection [193]. Assuming that the metasurface imposes the phase profile $\Phi(x, y)$, we assign the virtual wave vector $k_\Phi = \nabla\Phi_x\hat{x} + \nabla\Phi_y\hat{y}$ to the metasurface. In this context, the momentum conservation law for wave vectors can be expressed as

$$\begin{aligned}
k_i\sin\theta_i\cos\phi_i + \frac{d\Phi_x}{dx} &= k_r\sin\theta_r\cos\phi_r \\
k_i\sin\theta_i\sin\phi_i + \frac{d\Phi_y}{dy} &= k_r\sin\theta_r\sin\phi_r
\end{aligned} \tag{8.5}$$

where $\frac{d\Phi_x}{dx}$ and $\frac{d\Phi_y}{dy}$ describe the gradients in the x and y directions, respectively.

Since we can address any given (oblique) wave with a translation formulation [110], let us consider the normal incident wave case ($\theta_i = \phi_i = 0$) without loss of generality. Assuming air as the medium of the incident and reflected wave, we can simplify the formulation above as

$$d\Phi_x = \frac{2\pi dx \cos\phi_r \sin\theta_r}{\lambda_0}, \quad d\Phi_y = \frac{2\pi dy \sin\phi_r \sin\theta_r}{\lambda_0} \tag{8.6}$$

which express the change in phase (Φ_x and Φ_y) that needs to be performed per unit of distance (dx and dy) in the x and y directions. Then, we set the unit cell size ($d_x = d_y = D_u$) in Eq. (8.6) to obtain the phase required at the mn-th unit cell as

$$\Phi_{mn} = \frac{2\pi D_u(m \cos\phi_r \sin\theta_r + n \sin\phi_r \sin\theta_r)}{\lambda_0}. \tag{8.7}$$

To assign states to each unit cell, the required phase Φ_{mn} is calculated for all the unit cells. Then, a closest-neighbor mapping is done between the required phase and that provided by the different unit cell states. For instance, in our particular case where the number of states is $N_s = 4$, we have $\{s_0, s_1, s_2, s_3\}$ with the respective phases of $\{45, 135, 225, 315\}$ degrees, then required phases of $53°$ and $188°$ would be mapped to s_0 and s_2 states, respectively.

The aforementioned method provides a set of coding patterns that achieve the desired functionality. For the sake of exemplification, we provide the coding pattern of a particular metasurface two different reflected directions in Fig. 8.5 assuming normal incidence. It is observed that, depending on the reflection angle, the gradients on the x and y directions differ.

8.1.2.4 Performance Metrics

The metasurface model allows us to obtain the far-field pattern for a given HSF coding. This provides valuable, yet only qualitative insight on the response of the HSF. To quantify performance and extract behavioral trends, the use of a set of performance metrics is suggested. In this work, we consider the following metrics:

■ **Directivity** ($D(\theta, \phi)$)**:** The directivity describes concentration of energy at a given direction. We assess the directivity in the target angle of reflection $D(\theta_r, \phi_r)$ and the actual angle of reflection $D(\theta_a, \phi_a)$ as they are relevant to our analysis.

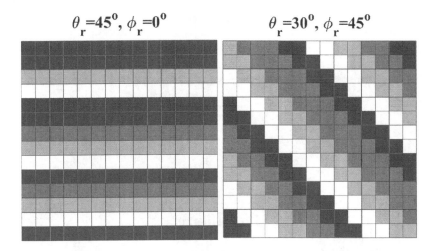

Figure 8.5 Coding of a square HSF with 15 unit cells per side and 4 unit cells states for different reflected angles. Each shade represents a different state with equispaced reflection coefficient phases.

- **Target deviation** (TD)**:** While our HSF aims to steer the beam to a particular angle, there may be a difference between the targeted angle and the actual reflected angle due to inaccuracies in the phase profile. We call this factor target deviation and calculate it as the Euclidean distance between the angle pointing to the direction of the reflected main lobe (θ_a, ϕ_a) and the desired angle for steering (θ_r, ϕ_r) as

$$TD = \sqrt{(\theta_r - \theta_a)^2 + (\phi_r - \phi_a)^2}. \qquad (8.8)$$

- **Secondary lobe level** (SLL)**:** The SLL is defined as the ratio (in dB) of the far field strength in the direction of the side-lobe nearest to the main beam to the far field strength of the main beam.

- **Half power beam width** $(HPBW)$**:** The waist of the main reflected beam defines the resolution of steering. The HPBW, that calculates the beam width at the -3dB of a normalized lobe, is a conventional factor to assess the beam width.

8.1.2.5 *Validation*

Before delving into the scalability analysis, we first assess the accuracy of the proposed method through a comparison with full-wave simula-

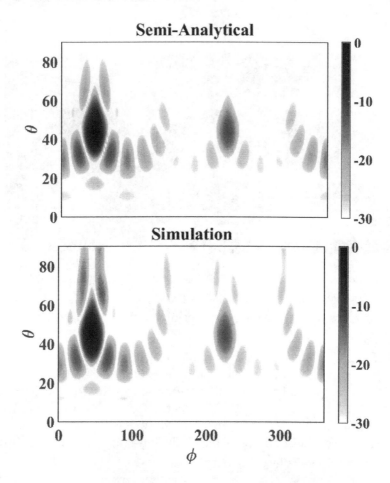

Figure 8.6 Normalized power radiation in dB of an HSF with $D_u = \lambda/3$ and $D_m = 5\lambda$ targeting $\theta_r = \varphi_r = \pi/4$ calculated by means of our analytical methodology (top) and through numerical methods (bottom).

tions. We simulated an HSF with $D_u = \lambda/3$ and $D_m = 5\lambda$. The HSF is coded to point the main lobe of radiation towards $\theta_r = \varphi_r = \pi/4$.

As shown in Figure 8.6, the semi-analytical method is in close agreement with the simulation and both show how radiation is mostly directed towards $\theta_r = \varphi_r = \pi/4$. The simulation time is reduced by approximately three orders of magnitude, thus exemplifying how analytical methods enable our scalability analysis approach.

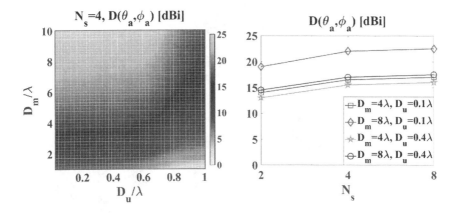

Figure 8.7 Left: Directivity at the direction of maximum radiation $D(\theta_a, \varphi_a)$ as a function of the size of unit cells and metasurface for $N_s = 4$ (left). Right: Directivity at the direction of maximum radiation $D(\theta_a, \varphi_a)$ as a function of the number of unit cell states for four different design points. In all cases, the HSF targets $\phi_r = \theta_r = \pi/4$ as direction of reflection.

8.1.3 Performance Scalability

This section lays out the results of the scalability analysis. We consider the metrics described above and evaluate them for a wide variety of design points encompassing multiple unit cell sizes D_u from 0.01λ to λ and metasurface size D_m from λ to 10λ. Additionally, we take several points of the design space and repeat the analysis scaling the number of states N_s from 2 (single-bit HSF) to 8 (three-bit HSF). Sweeping the parameters by an order of magnitude allows to identify the frontier between relevant and irrelevant design spaces. In all cases, the reported results are for a particular target angle $\varphi_r = \theta_r = \pi/4$ and for normal incidence. In future works, we will perform a scanning analysis aiming to evaluate how incidence and reflection angles impact the performance of differently sized HSFs.

8.1.3.1 Directivity

We start the investigation with the evaluation of the directivity as an indicator of power density in the direction of the main lobe. We use the maximum directivity as normalization factor in all plots. Figure 8.7 plots the directivity as a function of D_u and D_m and for three N_s values. It is clearly observed that the directivity increases with the

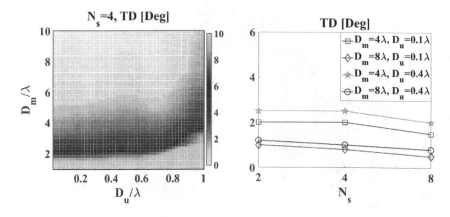

Figure 8.8 Left: Target Deviation (TD) as a function of the size of unit cells and metasurface for $N_s = 4$ (left). Right: TD as a function of the number of unit cell states for four different design points. In all cases, the HSF targets $\phi_r = \theta_r = \pi/4$ as direction of reflection.

MS size D_m: at $N_s = 4$, it raises 15 dB from λ to 3λ, whereas it increases another 10 dB from 3λ to 10λ. The effect of the unit cell density (inverse of D_u) is only apparent above $D_u = \lambda/2$. Below that value, the directivity becomes independent on the size of the unit cell –thus discouraging the use of very small unit cells due to the associated increase in complexity and cost.

The impact of increasing the number of states is noticeable from $N_s = 2$ to $N_s = 4$, with an improvement of $\sim$3 dB in most points. This is because, for $N_s = 2$, the reflected wave is actually split into two identical lobes directed to two symmetrical angles, which explains the 3 dB difference. Beyond that, adding a third coding bit does not have a significant impact on the directivity of the main beam.

8.1.3.2 Target Deviation

The TD needs to be carefully evaluated since small changes in the pointing angle can drastically affect the beam steering performance. Figure 8.8 reveals the trends of the TD versus dimensional parameters normalized to the wavelength.

The plots show that TD is mostly affected by the unit cell and metasurface dimensions. Decreasing the size of the unit cells reduces the pointing error of the MS because the spatial resolution is improved and therefore, the discretization error is reduced. The dependence of the TD with respect to the number of states follows a similar reasoning

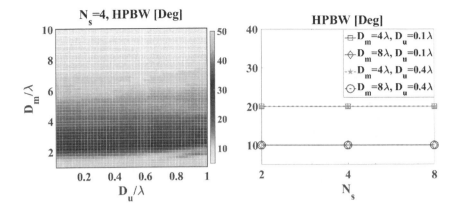

Figure 8.9 Left: Half Power Beam Width (HPBW) as a function of the size of unit cells and metasurface for $N_s = 4$ (left). Right: HPBW as a function of the number of unit cell states for four different design points. In all cases, the HSF targets $\phi_r = \theta_r = \pi/4$ as direction of reflection.

than for the unit cell size, but in this case the phase quantization error is minimized rather than the spatial discretization error. However, unlike in the case of the directivity, the improvement is hardly noticeable for the design points here evaluated.

8.1.3.3 Half Power Beam Width

Spatial resolution is a significant quality factor of scanning devices as it determines the minimum distance of two points in angular space that can be discriminated by the scanner. The resolution is inversely proportional to the beam width and $HPBW$ is a well-known parameter for beam width estimation. Figure 8.9 plots $HPBW$ as function of the dimensional parameters and the number of states. We basically observe that, as expected, the HSF size is the main determinant of the beam width. The main reason is that the aperture of the device increases. We achieve a steady beam width below 15° for $D_m \geq 6\lambda$ approximately. The lowest value achieved within the bounds of our exploration is around 5°. The impact of the discretization and quantization error, given by the unit cell size and the number of states, is generally irrelevant in this case.

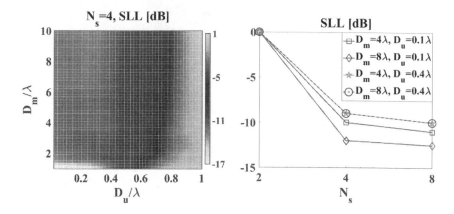

Figure 8.10 Left: Side-Lobel Level (SLL) as a function of the size of unit cells and metasurface for $N_s = 4$ (left). Right: SLL as a function of the number of unit cell states for four different design points. In all cases, the HSF targets $\phi_r = \theta_r = \pi/4$ as direction of reflection.

8.1.3.4 Side Lobe Level

The evaluation of the SLL is especially relevant as specific applications are constrained in power and aim for high efficiency. In this regard, the SLL is important as it provides a first-order estimation of the power that may not reach the target. Figure 8.10 shows the SLL versus dimensional parameters normalized to the wavelength. As previously mentioned, $N_s = 2$ is a particular case where the scattered field is split into two identical beams in symmetric directions, meaning that a constant value of $SLL = 0$ dB would be associated with the secondary beam throughout the design space.

The results shown herein demonstrate that the size of the unit cell is the key factor for the improvement of SLL. It is clearly observed that $D_u = \lambda/2$ marks a frontier that separates a region with good SLL below -12 dB from another region with SLL in excess of -9 dB. It is also worth remarking that the SLL keeps improving as we introduce a third bit of coding ($N_s = 8$). This reinforces the intuition that the SLL is mainly affected by errors in the discretization and quantization of the space-phase.

8.1.4 The HyperSurface Energy Footprint, Cost, and Performance

The results in the previous section have confirmed that large MS with little or no discretization error (unit cell size tending to zero) and phase quantization error (large number of unit cell states) consistently yield the best performance for beam steering. In fact, physical size and resolution appear to be more important than the phase. These remarks are important and provide insight on the optimal design points *if we just care about performance.* Therefore, the results above do not provide a unified design guideline (especially if we start to consider cost and complexity as aspects preventing unlimited scaling).

While user requirements indicate the acceptable thresholds of the performance metrics, fabrication restrictions and operational limitations (e.g., available power, space and other overheads) bound our design space [194]. Therefore, the design of an HSF should be tackled from a combined performance–cost perspective so as to deliver an effective and efficient platform for electromagnetic manipulation. In this section, we provide a qualitative discussion of how cost metrics could impact the design of an HSF.

To illustrate our case, first, we show a graphical representation of the HSF structure in Figure 8.11. Essentially, the HSF receives external programmatic commands from a gateway controller that are disseminated to the internal control logic at the controller chips via chip-to-chip interconnects and routing logic [137, 140]. These commands contain the state (within the discrete set of possible states) that should be applied to each unit cell. The control logic translates the state into an analog value to be applied to the tuning element, e.g., the voltage applied to a varactor to achieve a target capacitance. Additionally, embedded sensors can pick up data from the environment and send it to the control logic or external devices again via the communications plane.

8.1.4.1 Cost and Power Models

Clearly, the addition of sensors, actuators, and other internal circuitry will impact the power consumption and fabrication cost of the device. For the sake of exemplification, let us assume that we have an HSF of $N \times M$ unit cells serviced by $n \times m$ chips. In a column, each chip of size D_c gives service to $X = M/m$ unit cells of size D_u. The relation between M and m should be obtained based on D_c and D_u. A good approximation would be that a chip of size $D_c = D_u$ (even smaller, perhaps $D_u/2$) can be placed at the center of a cluster of 2×2 unit cells. A chip of size $D_c = 2D_u$ (even smaller, perhaps $3D_u/2$) can be placed at the center of a cluster of 3×3 unit cells.

An upper limit for the size of D_c would be 20–30 mm due to the manufacturing constraints of chips nowadays related to the die yield

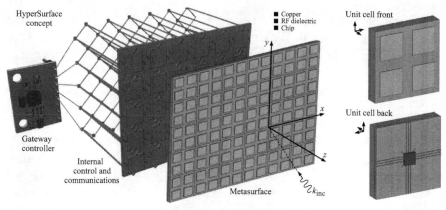

(a) Front, back, and layered representation [130]

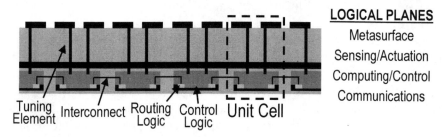

(b) Cross-section and logical planes

Figure 8.11 Graphical representation of a possible HSF implementation, which includes the metasurface plane with the metallic patches and the substrate, the sensing/actuation plane with the tuning elements and sensors, the computing/control plane containing the controller chips, and the communications plane containing the routing logic and interconnects. A gateway controller interfaces the HSF with the external world. From [155].

[73]. The size of D_c, on the other hand, has also a lower limit based on the technology and on the functionality that is attributed to it within an HSF. In the prototype case study outlined in this book, the chip should be large enough to host 24 pins to (1) power the chip and (2) to provide a connection to neighbouring unit cells. The size of a pin depends on the technology. Thus, in summary, the cost of the HSF has two main components:

■ **Technology:** Designers may want to accommodate as many pins as possible per chip so as to improve the intra-HSF bandwidth, even in the case that tiny chips are required to control high-frequency unit cells. In general, however, newer technologies allowing for smaller form factors are more expensive.

■ **Size:** Bigger chips are more expensive as they occupy a larger fraction of the fabrication die and have the potential to host more transistors/pins. Let us leave this term as a parameter, but also let us assume a linear increase of cost when we increase the size.

The cost of the MS and integration will probably depend on M or D_u in the first case, and on m in the latter case. In particular, we assume

$$Cost = C_{chips} + C_{meta} + C_{int} = CH(C_T, C_S) + C_{meta} + C_{int}, \quad (8.9)$$

where C_{chips} is the cost of chips, C_{meta} is the cost of the metasurface layer, and C_{int} is the cost of integrating the different sub-systems together. Within the first term, C_T is a technology factor and C_S is a size factor determining the cost of a single chip, whereas CH is the number of chips to be integrated within the HSF. Here, $CH = n \cdot m$, where n and m are the number of chips in the x and y directions of the HSF plane.

The energy consumption of the HSF, as any other circuit or processor, has two components. First, the static power that the controller chips consume just for being connected to a power source due to, for instance transistor leakage, and that does not depend on the workload. Second, the power that is consumed dynamically as a result of the switching of transistors and the passing of current. Thus, the total power P is divided between static power P_{sta} and dynamic power P_{dyn},

$$P = P_{sta} + P_{dyn}. \quad (8.10)$$

On the one hand, the static power mainly depends on the technology node, i.e., smaller transistors in newer technologies suffer from higher leaking currents, as well as on the number of chips. Further, if integrated components such as the varactors continually drain current, the static power will be dependent on the number of unit cells. Some models of static power exist in the literature [153], but a deeper analysis and, ideally, measurements from existing prototypes [187], will be required to have an accurate approximation of the static power consumption of an HSF.

On the other hand, the dynamic power is basically how many current-draining actions are taken per second multiplied by the energy consumed by each of these actions. An example is the switching of transistors from OFF state to ON state, which generates a transient current. We could consider, as an approximation, modeling the power consumption required to change the state of a single unit cell E_{chg}. Then, the dynamic power would be

$$P_{dyn} = \alpha \times f \times N^2 \times E_{chg}, \qquad (8.11)$$

where $\alpha \in (0, 1]$ is the probability of a unit cell changing state, f is the expected speed of state changes over time, which is application-dependent, and N^2 stands for the number of unit cells in the HSF. The energy consumed to change the state of a single unit cell, on its turn, depends on i) the energy consumed to send the message from gateway to unit cell, which depends on the number of routers traversed within the HSF as discussed in Section 8.2, and ii) the energy consumed for the controller to compute/apply the new state. These will in turn depend on the number of chips (each chip has one router). The traffic analysis can help to extract some approximate values of α, f, and the average number of router traversals from the gateway to the unit cell.

8.1.4.2 Application-Specific Figures of Merit

An additional point worth making is that the present analysis does not take the application requirements into consideration. For instance, it is a well-known problem that, although narrow beams provide high efficiency (and may be in fact necessary in scenarios such as terahertz wireless communications [18]), slight target deviations can lead to loss of connectivity. On the contrary, wider beams are less efficient, but also less prone to disruption.

For all these reasons, here we propose a figure of merit that attempts to put the different performance metrics together and introduce user requirements as well. The figure of merit is defined as

$$FoM_1 = D(\theta_r, \phi_r) - SLL - \frac{TD}{HPBW} - \frac{|Aperture - HPBW|}{HPBW} \qquad (8.12)$$

where $Aperture$ is an arbitrary beam width set as a specific requirement by the user/application. To incorporate the multiple performance metrics, we equalize the units using normalized values. D and SLL are converted to percentages, whereas we model the tradeoff between beam width and accuracy by dividing TD by $HPBW$. This way, the importance of the TD value increases for narrow beams, which are the most vulnerable against connectivity issues. The last term models how close is the MS to achieving the specific requirement of beam width set by the user or application. Note that, with all these considerations, the range of the figure of merit is $FoM \in [0,1]$ and a high value is preferred. Finally, we also note that weights can be applied to the different terms to create performance profiles according to the requirements of different applications. However, we leave such analysis for future work.

Figure 8.12 plots the figure of merit as a function of the dimensional parameters with $N_s = 4$ (which provides a good balance between performance and cost). We repeat the plots for different values of *Aperture* (10°, 20°, 40°) illustrating cases of high, average, and low directionality requirements. At a first glance, the MS size seems to play a bigger role in determining overall performance. For narrow beam applications, the figure of merit points to large and fine-grained MS as a necessity; whereas smaller and coarser-grained designs can be affordable as the beam requirements are relaxed.

8.1.4.3 Performance-Cost Analysis

To bridge this gap, the methodology presented in this chapter can be combined with parametrized models accounting for the power consumption or cost associated to the integrated controllers and circuitry. This would allow architects to evaluate performance-cost tradeoffs with performance-cost figures of merit and, by adding weights to each metric, to find optimal design spaces.

To exemplify this, let us take the example of Section 8.1.4.2 and assume a primitive model that states that power or cost of the HSF scale linearly with the number of unit cells per dimension. This assumption is based on the fact that more unit cells imply the use of more controllers and higher transit of messages within the MS [137,140]. In our particular example, we divided FoM_1 above by the number of unit cells per dimension and normalized the result. The outcome, that we refer to as FoM_2, is shown in Figure 8.13. Intuitively, the tendency is to favor configurations with fewer unit cells within the range that yields acceptable performance.

8.2 THE HYPERSURFACE DATA TRAFFIC AS A SCALING CONCERN

As stated throughout this book, HSFs integrate a network of controllers within the structure of the metasurface. This embedded sub-system provides means for distributed intelligence for metasurface reconfigu-

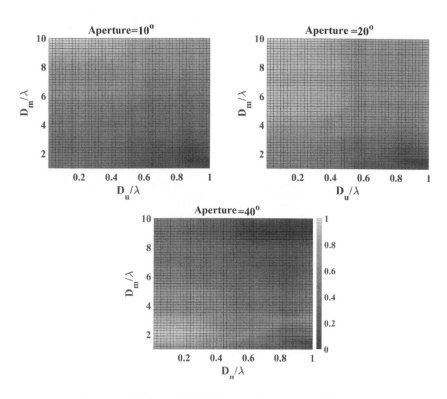

Figure 8.12   Figure of Merit (FoM$_1$) as a function of dimensional parameters for a 2-bit programmable metasurface targeting $\phi_r = \theta_r = \pi/4$ for three different values of targeted beam width *Aperture*. The values bar is common to all figures.

ration and control, while also being a primary source of cost and power consumption. In essence, controllers drive the reconfigurable unit cells and exchange information with neighbouring controllers using the integrated network.

To approach the design of the internal network of controllers properly, it is necessary to understand the workload that the internal HSF network will have to serve [163]. In other words, we need to characterize the expected traffic between the HSF's controllers in order to make informed design decisions regarding the HSF network. Due to the novelty of the programmable MS in general and of the HSF concept in particular, such an analysis has not been carried out in depth thus far.

This section is based on [137] and aims to bridge this gap by providing a traffic analysis of HSFs. Given the breadth of the problem, we first focus on a particular yet relevant functionality: anomalous reflection for

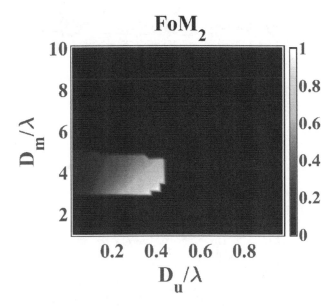

Figure 8.13 Performance-cost figure of merit. This is obtained from FoM_1 with *Aperture* $= 30°$ and assuming a power-cost model that scales linearly with the number of unit cells. Values close to 0 (1) refer to invalid (optimal) design points.

beam steering. To this end, we apply the semi-analytical methodology employed in the scalability analysis of Section 8.1 to obtain the state of each unit cell as a function of several design parameters (e.g., unit cell size) and application parameters (e.g., steering angle). We extend this methodology to enable the study of the variations in the unit cells' states as functions of changes in the required reflection angle. Assuming that every state change implies sending a message, as detailed in Section 8.2.1, these variations provide spatiotemporal information on the HSF traffic. By mapping angular changes to physical movements in real-world scenarios, the methodology summarized in Section 8.2.2 can be applied to real use cases such as tracking a mobile user in a 5G network. We first employ the developed methodology to investigate the traffic patterns for a set of simple yet representative move types in Section 8.1.3. We then study the traffic generated in a more complex indoor mobility scenario with multiple surfaces deployed, as detailed in Section 8.2.4.

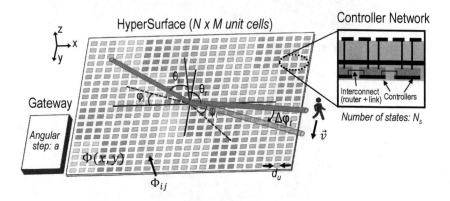

Figure 8.14 System model. Target moves with speed $\vec{v}$ changing the required reflected angle. An HSF with $N \times M$ unit cells of size D_u implements a phase profile $\Phi(x, y)$ to obtain the desired reflected angle. The phase of each unit cell Φ_{mn} is approximated to the closest phase among the N_s available states. The gateway reconfigures Φ_{ij} when angles vary more than the angular step a, which is a design parameter.

8.2.1 System Model

This section outlines the notation and main assumptions. Let us consider an HSF with $M \times N$ unit cells and a gateway as shown in Figure 8.14. Following the same notation as in Section 8.1.2, let us assume that each unit cell is squared with side D_u and can be configured in any of its N_s states. In the case of Fig. 8.14, $N_s = 4$ with each state represented by a shade. The metasurface is fully illuminated by a plane wave coming from $\{\theta_i, \phi_i\}$ that is reflected towards $\{\theta_r, \phi_r\}$. The aim of the HSF is to adapt to changes in either direction, possibly due to movement of the emitter or receiver with respect to the HSF. This could model well applications where a given target needs to be tracked, such as communication with directive antennas in 5G environments.

The rest of the section is laid as follows. The types of movements assumed in the workload analysis are described in Section 8.2.1.1. Sections 8.2.1.2 and 8.2.1.3 outline the assumptions made with respect to the HSF gateway and network of controllers. For details on how the coding patterns of the HSF are obtained, we refer the reader to Section 8.1.2.

Figure 8.15 Mobility cases considered in this work.

8.2.1.1 Mobility Model

Without loss of generality, let us consider an HSF that is fixed onto a wall and is used to track the position of users or objects moving along a vector $\vec{v}$. This could model, for instance, a scenario where the objective is to avoid service disruption in wireless 5G networks with highly-directive antennas [94]. The position of the moving objects can be expressed in spherical coordinates using the HSF as point of reference, thus allowing to obtain the desired reflection angle over time. The trajectories under study are illustrated in Fig. 8.15 and described next.

- **Case** A**:** The target moves in a straight line parallel to the surface where the target is at the same height as the surface. Motion starts from a point far away from the surface and finishes when the object is directly in front of it. This scenario represents the movement of a mobile user, thus we assume a default speed of 1.4 m/s, which is the speed of walking of the average human.

- **Case** B**:** The target describes a projectile motion parallel to the surface starting from a point close to the surface and moving away from it, as illustrated in Fig. 8.16. This could represent a typical case of radar tracking. Unlike the horizontal movement described above, the projectile motion changes both the azimuth and elevation angles of the reflected signal. The initial speed is assumed to be 30 m/s.

- **Case** C**:** The target takes arbitrary leaps which results into abrupt changes in location as opposed to the gradual change in the aforementioned cases. This case represents a person moving in an area with multiple mobile obstacles resulting in intermittent connection with the surface. Using the same settings of Case A, we model the arbitrary movement by randomly changing the azimuth angle of the reflected signal.

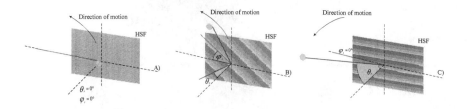

Figure 8.16 Illustration of tracking in Case B, which models a projectile movement parallel to the HSF.

In all cases, we consider that the metasurface is fixed onto a wall and normal incidence. The position of the moving objects can be expressed in spherical coordinates using the HSF as point of reference. The reflected angles are obtained accordingly. For each of the cases, thus, we can trace the position of the emitter and receiver with respect to the HSF and compute the required metasurface code in each case.

8.2.1.2 Gateway Model

For the purposes of traffic analysis, we will consider that the HSF is equipped with at least one gateway. The gateway is in charge of gathering the changes in incidence and reflection angles and reprogramming the HSF in order to adapt to those changes. In this work, we model the gateway operation as follows. Whenever any of the angles changes by more than a given pre-defined amount, which we define as *angular step a*, the gateway computes the new state for each unit cell. To save power, the gateway communicates only to the unit cells that need to be change the state. For this approach to be possible, we assume that the gateway:

1. has external sensing capabilities to obtain the angle of incidence $\{\theta_i, \phi_i\}$,

2. communicates with other external devices to obtain the targeted direction $\{\theta_r, \phi_r\}$,

3. can compute the phase profile Φ_{mn} and, thus, the state of each unit cell using the model from Section 8.1.2.

8.2.1.3 Embedded Controller Network

This work assumes a simple HSF design where each controller drives a single unit cell. The controllers are interconnected among them and

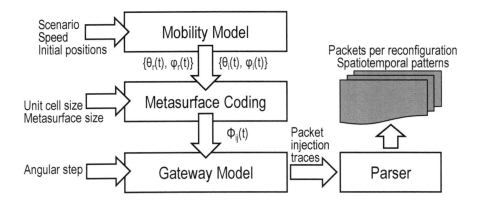

Figure 8.17 Summary of the evaluation methodology.

to the gateway via an internal network, with two purposes: (a) for the gateway to send the reconfiguration requests to the controllers, and (b) for the controllers to send acknowledgment packets to the gateway.

In order to model how fast does the HSF serve the signaling between the gateway and the controllers, it is necessary to model the chip interconnects. This section assumes the network topology and asynchronous communication means described in [84, 140].

8.2.2 Evaluation Methodology

Figure 8.17 illustrates our methodology to obtain the traffic within the metasurface. MATLAB scripts are employed to simulate the movement and to obtain the corresponding unit cell state matrices. More specifically, we first evaluate the incidence and reflection angles given the positions of the HSF, the illumination source, and the moving target (mobility model). Then, we use Eq. (8.6) to obtain the phase gradients and then Eq. (8.7) to calculate the phase Φ_{mn} of each unit cell. Finally, we take the unit cell state that yields the phase that is closer to Φ_{mn}, this is, we perform the metasurface coding.

The methods described above allow to obtain the matrix of unit cell states for any incidence and reflection conditions. Iterating over such calculations, we can obtain successive unit cell state matrices corresponding to a given movement with any angular granularity. To model the gateway, we only obtain the unit cell state matrices in steps corresponding to the *angular step* parameter. A *diff* operation between adjacent unit cell state matrices describes which unit cells need to be changed and, given our assumptions, which unit cells will receive packets from the gateway.

8.2.2.1 Relevant Inputs

For each type of movement, we collect traces that describe the time instant at which packets are generated as well as their intended destinations. The traces are obtained and classified as a function of the following parameters:

1. **Metasurface size** $N \times M$**:** The number of unit cells has an impact on the absolute amount of transmitted messages. Observing the impact of the metasurface size on both the required throughput can give design guidelines for HSFs. By default, we take $M = N = 50$.

2. **Number of states** N_s**:** The calculations used to assign cell states rounds to the state that yields the phase which is closer to Φ_{ij}. Higher number of states corresponds to a finer resolution phase gradient that improves the metasurface performance, but possibly at the cost of transmitting more packets. By default, we set $N_s = 4$.

3. **Angular step** a**:** Some applications may require tracking objects at very fine-grained angular resolution. In that case, the gateway would probably need to trigger state changes more often. This is modeled via the angular step, whose default value is set to $a = 5°$.

8.2.2.2 Traffic Analysis Metrics

The traffic traces are parsed to obtain the following relevant metrics:

1. **Percentage of state-changing cells:** The comparison between successive HSF state matrices determines the percentage of cells that must be adjusted to accommodate the change in the position of the target or the illumination source.

2. **Destination matrix:** The traffic analysis yields a matrix containing the ratio of packets delivered to a given destination with respect to all the transmitted messages.

3. **Reconfiguration delay:** Finally, we are capable of obtaining the delay of distributing the state-changing messages by plugging the traces to an AnyLogic-based custom-made simulator. On the simulator, the Manhattan-like topology described in [128, 140] is built. The reconfiguration delay refers to the time between the first message transmission to the last message reception.

8.2.2.3 Walkthrough Example

For the sake of exemplification, consider tracking an object moving as in Case B. Figure 8.16 shows snippets of the surface states during the full range of the movement (each shade represents a distinct unit cell state). The starting point is in Fig. 8.16A, where the elevation and azimuth angles of the reflected wave are $\theta_r = \phi_r = 0°$. This is also the initial state of the unit cells. The object starts moving in the direction shown in the figure with initial velocity of 30 m/s. When the location of the object changes by more than the angular step a in either the azimuth or elevation planes, the gateway is triggered. From a local data base, the gateway gathers the commands that reconfigure the surface such that the reflected wave arrives to the target at the new location. The commands are then injected into the controller network and forwarded to the individual controllers. Note here that the reconfiguration commands are sent only to the cells that change state. This process is repeated every time the tracked object changes its location by more than a. This process thus generates traces that can be later fed to the AnyLogic simulator to obtain the network delay.

8.2.3 Workload Characterization

In this section, we employ the methodology proposed above to examine the HSF workload. We start by analyzing the spatio-temporal characteristics of the traffic in Section 8.2.3.1. Then, we examine the effect of varying the parameters of the surface on the network delay and the traffic generated by the different motion scenarios in Sections 8.2.3.2 and 8.2.3.3, respectively.

8.2.3.1 Spatio-Temporal Intensity

Here, we analyze the spatio-temporal intensity and rate of updates generated by different types of movements. This is accomplished by showing how often reconfiguration commands are injected to the system. We use this to assess the rate of injection throughout the tracking

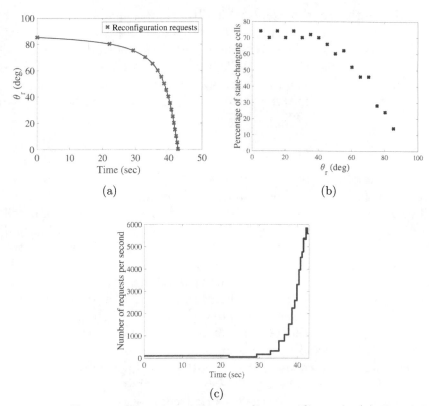

Figure 8.18 For an object moving according to Case A: (a) Reconfiguration requests during the tracking of the object. (b) Percentage of reconfigured cells versus the azimuth angle of the reflected signal. (c) The number of reconfiguration requests per second.

process. In addition, we visualize the spatial distribution of the required configuration through heat maps that correspond to the surface.

The generation of reconfiguration requests relies on the location of the tracked object, the motion pattern and the angular step. While the HSF can only sense the change in the incident and reflection angles, the latter is an outcome of the distance and the height difference between the tracked object and the surface, and the motion pattern. This is depicted in Fig. 8.18(a) where a motion of Case A is tracked. The markers indicate the time instances when reconfiguration requests are sent from the gateway to the system. Figure 8.18(a) shows that reconfigurations are more frequent as the object moves closer to the surface. For example, more than 88% of the reconfigurations are required within the last third of the motion. This is expected since the reflection angle changes faster when the object is closer to the surface.

Note that not all cells are reconfigured at each request. Rather, a portion of the cells is reconfigured. From Fig. 8.18(b), it is clear that the percentage of reconfigured cells is highly dependent on the preceding and the currently targeted angles. For example, when the tracked object moves from $\theta_r = 25°$ to $\theta_r = 20°$, 70% of the cells are reconfigured. When the change is from $\theta_r = 80°$ to $\theta_r = 75°$, on the other hand, only 28% of the cells are reconfigured in spite of the fact that the change in both cases was 5°. It is worth mentioning here that the initial configuration is excluded from the graph as it does not contribute in showing the trend of change.

When a change in the state of the HSF is required, reconfiguration requests are streamed into the system, which renders the injection inherently bursty (i.e., a burst of reconfiguration commands are injected every time a proper change in the angle occurs). However, the number of reconfiguration requests and the frequency with which the requests are made determine the injection rate. In Fig. 8.18(c), we show how the injection rate significantly increases when the changes in the reflection angle are more frequent. Coincidentally, frequent changes in Case A occur for target angles affecting a higher percentage of unit cells, pushing the injection rate further. This is an important result because excessive injection rates can be an offset of congestion within the controller network.

Another way to evaluate traffic is the spatial distribution of cells to be updated. To visualize this we resort to heat maps, where hotter spots represent the regions of the HSF where controllers receive higher numbers of packets. Figure 8.19 depicts the heat maps corresponding to the three scenarios of movements. Figure 8.19(a) shows the case of gradual change in θ_r where the tracked object follows the motion of Case A. Figure 8.19(c), however, shows the case of arbitrary changes in θ_r where the object makes sudden unpredictable leaps, i.e., Case C. The heat maps demonstrate that the traffic is almost evenly distributed over the surface for arbitrary changes. In addition, if compared with the heat maps in Fig. 8.19(b), one can observe the difference in the traffic distribution when the elevation angle of the reflected wave is fixed to 0 (i.e., the HSF is at the same height of the tracked object) and when it is variable. This information can be used in designing the routing mechanism, congestion control techniques and in placing the HSF tiles.

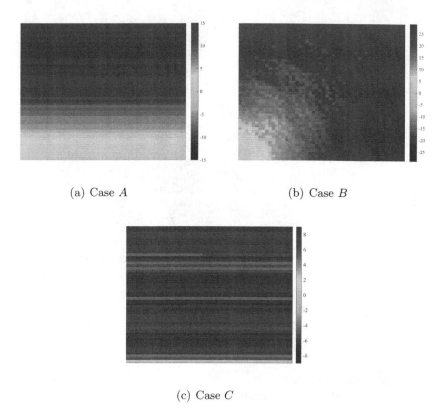

(a) Case A (b) Case B

(c) Case C

Figure 8.19   Spatial distribution of traffic for the three considered movement scenarios.

8.2.3.2   Reconfiguration Delay

The effect of changing the number of states on the delay is shown in Fig. 8.20. Since a higher value of N_s produces larger traffic, it is natural for the network delay to increase as observed in the figure.

Furthermore, we investigate the effect of changing the size of the surface on the relationship between the percentage of the reconfigured cells and the delay as shown in Fig. 8.21. We observe that the linear relationship between the percentage of reconfigured cells and the delay holds for different sizes of the surface. Although varying the size creates a discrepancy in the slopes of the graphs, yet this does not impact the linearity of the relationship. The figure also shows that larger surfaces achieve higher delays which is expected since packets must travel longer paths to change the states of different unit cells.

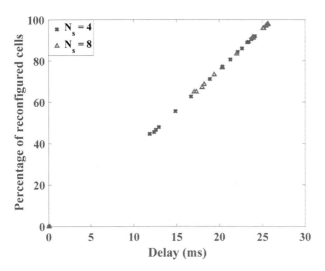

Figure 8.20 Percentage of reconfigured cells vs delay for two values of N_s.

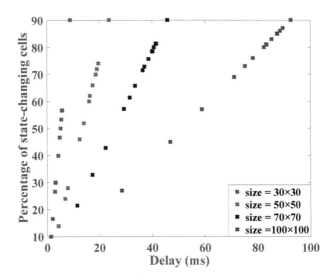

Figure 8.21 The linear relationship between the percentage of reconfigured cells and the delay for different sizes of the surface.

8.2.3.3 Sensitivity Analysis

Here, we change some of the metasurface parameters, namely the angular step and the number of states and observe the effects on the

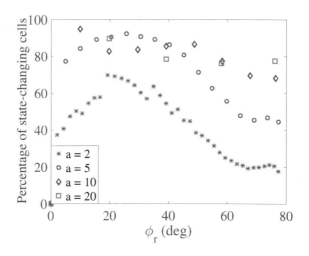

Figure 8.22 Elevation angle of the reflected signal versus the percentage of state-changing cells for different values of the angular step for projectile motion.

performance. In the previous results, the angular step was set to $5°$. The value of a can be used as an indication of the beam width, such that smaller values of a infer narrower beam widths and higher tracking precision requirements. We investigate the impact of changing a when an object moving according to Case B is tracked. Case B implies that both the azimuth and elevation angles are varied throughout the movement. A reconfiguration is requested every time either of the two angles change by $a°$.

Figure 8.22 depicts the percentage of state-changing cells corresponding to different values of the elevation angle of the reflected signal for different values of a. An angular step of $2°$ achieves the smallest percentage of change, and thus traffic, over the entire range of angle variation, which suggests that the HSF may operate faster. However, a small angular step requires a narrow beam which might increase the complexity of the surface fabrication. In addition, as the value of a decreases the rate of updates is expected to increase. The effects of this increase on the injection rate will be investigated in the future. It is worth mentioning here that the movement in Fig. 8.22 is projectile movement. However, for clarity we only show the change in angles in the first half of the motion (until the object reaches the highest point and before dropping).

Figure 8.23 Spatial distribution of traffic in the case of projectile motion for values of N_s of $N_s = 4$ (left), $N_s = 8$ (middle) and $N_s = 16$ (right).

The angular step, the smallest change in the angle sensed by the surface, is strongly correlated with the beam width. For instance, a narrow beam implies a small value of a and thus higher tracking resolution. In Fig. 8.22, we plot the percentage of updated cells as a function of the elevation angle for different values of a in the case of projectile movement. For clarity, we consider half of the time span of the entire motion range. Note that in this type of motion the azimuth and elevation angles both change as the object moves, thus, a reconfiguration is required whenever either angle changes by $a°$. The figure shows that smaller values of a demand more frequent reconfigurations but with a smaller number of reconfigured unit cells at each reconfiguration. While this implies lighter overall traffic and hence a faster operating HSF, it can complicate the surface fabrication due to the narrow beam required for small values of a.

Figure 8.23, on the other hand, shows the heat maps produced from tracking an object in the projectile movement for different numbers of states of the HSF cells, namely $N_s = 4, 8, 16$. As the number of states increases, the traffic becomes heavier over the entire surface, while for smaller number of states the traffic is lighter especially at the bottom-left corner. This might stem from the fact that the bottom-left corner is where the programming starts (initial state is always zero phase). In rather large phase gradients, that area remains largely at the same state.

8.2.4 Indoor Mobility Scenario

In an indoor scenario, HSFs are envisioned to coat objects like walls and furniture [94], in which case multiple surfaces can be responsible for the routing configuration of the Non-Line of Sight (NLoS) path between two users. In this section, we consider such a scenario to investigate the generated traffic in a more realistic setting. As such, we characterize

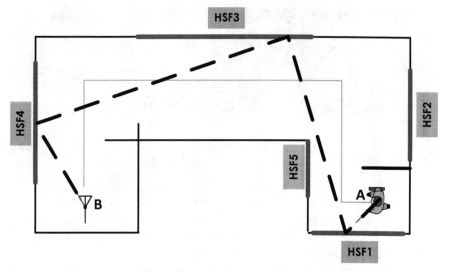

Figure 8.24 Indoor mobility scenario.

the traffic workload on the controller network of five MSs placed in an indoor environment as shown in Fig. 8.24. We assume that a mobile user moves from point A to point B following the trace indicated by the dashed line in the floorplan shown in the figure. In addition, the user is connected to the stationary access point at point B through the NLoS path created via the MS on the walls. The red lines represent the HSF tiles. The user is assumed to emit a very narrow beam at an angle of 55° to the left of the trajectory of the movement. The dimensions of the floor plan and the HSF tiles, and the trajectory of the movement of the user are all fed to an AnyLogic based simulator. The simulator produces the traces of the wave emitted from the user's mobile device until it reaches point B along with the incident and reflection angles of the wave at each HSF. It is important to note that the wave does not necessarily activate all the surfaces in the room. In the example wave trace shown in Fig. 8.24 (in the black dashed line) for instance, the wave does not activate $HSF2$ and $HSF5$. The path of the signal changes as the user changes locations.

The incidence and reflection information at every surface is then treated the same way as in the previous experiments to generate the spatial distribution of traffic and the rate of injection of the reconfiguration requests. Figure 8.25 shows the spatial distribution of traffic within the four tiles. It can be observed that $HSF4$ is used more than the rest of the surfaces in the considered scenario, whereas $HSF1$ is the least used surface. This is due to the fact that the user at point B

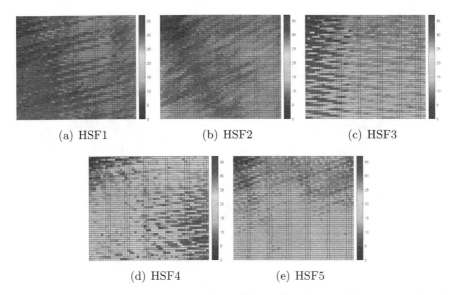

(a) HSF1　　　　　(b) HSF2　　　　　(c) HSF3

(d) HSF4　　　　　(e) HSF5

Figure 8.25　Spatial distribution of traffic in the five surfaces considered in the indoor mobility scenario.

is stationary and because of the narrow beam emitted at a 55° angle, the user can either be reached directly, i.e., through a LOS path or via HSF4. It is important to note that in reality a wider beam is emitted from the user's device, which is usually simulated as a bundle of narrow beams. Each beam has an additive effect to the incurred traffic. Hence, for ease of exposition we only consider one spatial channel of the beam, represented by a narrow beam emitted at angle 55°.

The rate of injection at each surface is shown in Fig. 8.26. The motion starts at $time = 10sec$ and lasts for about 14 seconds. Throughout the considered motion, the user loses connectivity for some time, as for example between the $14th$ and $20th$ seconds as one can indicate in Fig. 8.26(d). This loss of connectivity occurs due to the narrow beam assumed for the purposes of this simulation. Figure 8.26 shows that the rate of injection does not exceed 3 Mbps for all surfaces. However, due to the location of the tiles, different loads are on each. This information can be useful in designing congestion control algorithms and in placing the tiles.

8.3　CONCLUSIONS

This chapter has provided a metasurface-centric and application-oriented analysis of the prospective capabilities of future HSFs. We

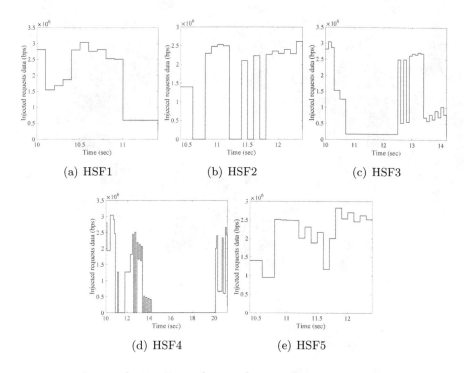

(a) HSF1 (b) HSF2 (c) HSF3

(d) HSF4 (e) HSF5

Figure 8.26 Rate of injection of reconfiguration requests in bps for the five surfaces considered in the indoor mobility scenario.

have focused on the link between the application requirements (represented as a set of performance metric values) and the metasurface essential characteristics (summarized in a small set of design input parameters) for the case of beam steering. In overall, we have argued and illustrated how this approach allows to dimension the metasurface and navigate the tradeoffs faced in the design process of the complete architecture, in the pathway to deriving useful guidelines of design for future intra-HSF networks.

In Section 8.1, we have presented and applied a methodology to study the scaling trends of HSFs by the order of magnitude. We have concluded that four unit cell states (2 bits) are enough to provide acceptable performance across a wide range of metrics. We also confirmed that large metasurfaces with very small unit cells provide the best performance in most metrics, although it is arguably the most costly solution from a fabrication perspective. Further, to provide unified design guidelines, we proposed performance figures of merit with which we illustrated which design regions can provide close-to-optimal results while minimizing potential cost. Future works could further advance in this aspect by incorporating accurate power and cost models that will allow us to determine the region of optimal solutions in terms of the performance-to-cost ratio, among other explorations.

In Section 8.2, we leveraged the analytical formulations of beam steering coding and, through the definition of typical movements in user-tracking applications and other assumptions, extracted a set of intra-HSF traffic traces. This allows not only to estimate the average communication load, but also to characterize the spatiotemporal features of such load. We have observed that both the amount of state-changing cells and the temporal distribution of requests clearly depend on the instantaneous reflection angle. Moreover, the specific trajectory of the user, understood as the succession of changes in the reflection angle, determines the spatial distribution of unit cell state variations. If the application requires fine-grained tracking, either in time or space, we can expect increases in the amount of traffic to handle through increases of the associated design parameters of the HSF. A further analysis performed in this chapter used traffic traces obtained in this characterization effort to evaluate the delay of prototype HSFs. We have seen, as expected, how the delay is dependent on the reflection angles as they involve more intense exchange of information.

CHAPTER 9

Applications of the Internet of Materials: Programmable Wireless Environments

Christos Liaskos, Ageliki Tsioliaridou, Sotiris Ioannidis

Foundation for Research and Technology Hellas, 71110, Heraklion, Crete, Greece

Shiai Nie

Georgia Tech, USA

Andreas Pitsillides, Ian Fuat Akyildiz

Computer Science Department, University of Cyprus

CONTENTS

9.1	Deterministic Wireless Propagation Control as a Concept		270
9.2	Modeling, Simulating, and Configuring PWEs—A Ray-Routing Approach Based on Graph Theory		272
	9.2.1	General Modeling and Properties of HyperSurface Functions	272
	9.2.2	Specialized Modeling of Function Inputs/Outputs	274
	9.2.3	Modeling Core HyperSurface Functions	277
	9.2.4	A Graph Model for Simulating and Optimizing Programmable Wireless Environments	283
	9.2.5	Modeling Connectivity Objectives	286
		9.2.5.1 Power Transfer Maximization	287
		9.2.5.2 QoS Optimization	288
		9.2.5.3 Eavesdropping Mitigation	289

9.2.5.4   Doppler Effect Mitigation 289
9.2.5.5   User Blocking 290
9.3   A K-Paths Approach for Multi-User Multi-Objective
Environment Configuration 290
9.4   Artificial Intelligence-Based Configuration of PWEs 294
9.4.1   Feed-Forward 296
9.4.2   Back-Propagation 296
9.5   The Novel PWE Potential in Communication Quality,
Cybersecurity, and Wireless Power Transfer 297
9.5.1   Multi-User Multi-Objective Showcase 300
9.5.2   Doppler Effect Mitigation Showcase 302
9.5.3   User Capacity and Stress Test 304
9.5.4   Evaluation of Neural Network-Based PWE
Heuristics 306
9.6   Conclusion 309

Disclaimer: This chapter adapts material from the following publications, with permission from the respective publisher:

■ Liaskos, C., Tsioliaridou, A., Nile, S., Pitsillides, A., Ioannidis, S., Akyildiz, I. (2019, July). An interpretable neural network for configuring programmable wireless environments. *2019 IEEE 20th International Workshop on Signal Processing Advances in Wireless Communications (SPAWC)* (pp. 1-5). IEEE.

■ Liaskos, C., Tsioliaridou, A., Nile, S., Pitsillides, A., Ioannidis, S., Akyildiz, I. F. (2019). On the network-layer modeling and configuration of programmable wireless environments. *IEEE/ACM Transactions on Networking*, 27(4), 1696-1713.

9.1 DETERMINISTIC WIRELESS PROPAGATION CONTROL AS A CONCEPT

The Internet of Materials can constitute an effective means for controlling the wireless propagation within a space, introducing programmable wireless environments (PWEs) [92, 103]. According to the PWE paradigm, planar objects–such as walls in a floorplan–receive a special coating that can sense impinging waves and actively modify them by applying an electromagnetic (EM) *function*. All metasurface-enabled wave modifications are possible, including altering the wave's direction, power, polarization and phase [89]. In this chapter we focus on the network-layer PWE configuration problem, i.e., which functions to deploy at the PWE coatings to serve a set of given user communication objectives within a space.

Apart from metasurfaces, coating technologies for PWEs include relays and phased antenna arrays [95]. Each technology comes with a range of supported functions, environmental applicability and efficiency degrees. Relays are 1 input–N output antenna pairs that can be placed over walls at regular intervals [179]. At each pair, one out of the N outputs can be selected, thereby redirecting the input wave in a partially customizable manner. Phased antenna arrays–also known as intelligent surfaces and reflect arrays [43, 152, 156]–are panels commonly comprising a number of patch antennas with half-wavelength size, in a 2D grid arrangement. At each patch, active elements such as PIN diodes are used for altering the phase of the reflected EM wave. Consistent wave steering and absorption is attained at the far field. Metasurfaces have $25 - 100^+$ times higher density of *meta-atoms*, allowing allows them to form any surface current distribution over them, thereby producing any EM output due to the Huygens principle [124]. Thus, highly efficient EM functions even in the near field can be attained. Moreover, as discussed in Chapter 4, HyperSurfaces come with the software programming interfaces that allows them to be treated as black-boxes, facilitating their direct integration into applications [99, 101], without knowledge of the underlying Physics.

A PWE is created by coating planar objects–such as walls and ceilings in an indoor environment–with *tiles*, i.e., rectangular panels of any aforementioned technology, with inter-networking capabilities [92]. The latter allow a central server to connect to any tile, get its state and set its EM function in an automated manner [93]. This maturity level reached at the physical layer of tiles opens a new research direction at the network level: *given a set of users with communication objectives within a PWE, what is the optimal EM function per tile to serve them?*

In this chapter we present a solution to this problem, able to handle multiple users, objectives and EM functions. User mobility, multiple objectives per user, multicast groups and partially coated PWEs are supported. The objectives include wireless power transfer and signal-to-interference maximization, as well as eavesdropping and Doppler effect mitigation. In order to achieve these traits, the present chapter details:

■ A systematic way of formulating and combining EM functions for applications.

■ The *EM profile* of tiles, a novel concept that describes the supported EM functions per tile and their efficiency.

■ A graph-based model to describe PWEs, and a way of transforming communication objectives to graph paths.

Extensive evaluations in multiple floorplans and topologies are presented as examples, yielding important conclusions about the maximum potential of PWEs and their user capacity in terms of maximal supported traffic load. Moreover, while the focus of the examples is wireless communications, the same algorithms can be applied for software-driven manipulation of wireless propagation in any setting.

9.2 MODELING, SIMULATING, AND CONFIGURING PWEs — A RAY-ROUTING APPROACH BASED ON GRAPH THEORY

This section provides an abstract model of the Physics behind metasurfaces, leading to a function-centric formulation of their capabilities. This formulation is then used for modeling PWEs as a graph, and describing its workflow and performance objectives as path finding problems.

Persistent notation is summarized in Table 9.1 for ease. (Notation used only locally in the text is omitted).

9.2.1 General Modeling and Properties of HyperSurface Functions

Let $\mathcal{H}$ denote the set of all HyperSurface tiles deployed within an environment, such as the floorplan of Fig. 9.1. A single tile will be denoted as $h \in \mathcal{H}$. Let $\mathcal{F}_h$ denote all possible EM functions that can be deployed to a tile h. A single function deployed to a tile will be $f_h \in \mathcal{F}_h$.

A function f_h is attained by setting the active elements of the HyperSurface accordingly. In this work we will assume that the correspondence between functions and active element states is known, and the reader is redirected to studies on *EM Compilers* for further details [99, 101].

Each function f_h receives a nominal input EM field, $\overrightarrow{E_{in}}$, (i.e., impinging upon the tile), and then returns a well-defined output $\overrightarrow{E_{out}}$ (i.e., a reflected, refracted, or no field–in case of perfect absorption), which can be abstracted as:

$$\overrightarrow{E_{out}} \leftarrow f_h \left(\overrightarrow{E_{in}} \right) \tag{9.1}$$

Table 9.1 Summary of Notation.

Symbol	Explanation
$\mathcal{H}$	The set of all tiles within an environment.
$h \in \mathcal{H}$	A single HyperSurface tile.
$\mathcal{F}_h$	The set of EM functions supported by a tile h.
$f_h \in \mathcal{F}_h$	A single function, deployed to tile h.
$f_h^{\text{ABS}}, f_h^{\text{STR}}, f_h^{\text{COL}}$	Absorption, Steering and Collimation functions.
$m_h^{\text{PHA}}, m_h^{\text{POL}}$	EM phase and polarization function modifiers.
$\overrightarrow{E_{in}}, \overrightarrow{E_{out}}$	Nominal input/output (EM field) of a function.
$\mathbb{I}, \mathbb{O} :$	EM function input/outputs as wave attributes:
$\left\langle \omega, \overline{D}, \mathbf{P}, \overrightarrow{\mathbf{J}}, \Phi \right\rangle$	¡frequency, direction, power, polarity, phase¿.
PW, FW	Subscripts denoting plain wave and focal wave.
g_h	Wave power gain/loss after impinging at tile h.
$\mathcal{G}\langle\{\mathcal{H},\mathcal{U}\},\{\mathcal{L}_h,\mathcal{L}_u\}\rangle$	Graph with tiles $\mathcal{H}$ and users $u \in \mathcal{U}$ as nodes, inter tile links $\mathcal{L}_h$ and user-to-tile links $\mathcal{L}_u$.
$\overrightarrow{p}_{n\,n'}$	A path in $\mathcal{G}$ as list of links from node n to n'.
$l_{n\,n'}$	A link in $\mathcal{G}$ from node n to n'.
$\text{Tx}\,(l_{uh}), \text{Rx}\,(l_{hu})$	Link labels denoting intended Tx and Rx users.
$\langle \ldots \rangle$	A tupple (group) of items.
$\{ \ldots \}$	A list of objects (single items or tupples).
$*$	Unintended (not nominal) type of quantity $*$.
$\| * \|$	The cardinality of a set $*$.

Consider the coordinate system over a tile, as shown in Fig. 9.2. In the most generic function case, $\overrightarrow{E_{in}}$ is defined over the $\phi = 90^o$ plane on the surface, while $\overrightarrow{E_{out}}$ contains the output field at any point $\{r, \theta, \phi\}$ around the tile. It is noted that a function f_h also defines the output to any, even *unintended input*, $\widetilde{E_{in}}$, which can exemplary arise when EM sources move, without adapting the tile functions accordingly. Therefore, relation (9.1) is generalized as:

$$\widetilde{E_{out}} \leftarrow f_h\left(\widetilde{E_{in}}\right) \tag{9.2}$$

We proceed to remark two important properties of the EM functions, stemming from physics:

EM functions f_h are symmetric [89, 174]:

$$\widetilde{E_{out}} \leftarrow f_h\left(\widetilde{E_{in}}\right) \Leftrightarrow \widetilde{E_{in}} \leftarrow f_h\left(\widetilde{E_{out}}\right) \tag{9.3}$$

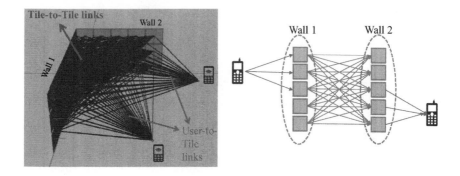

Figure 9.1 3D and 2D illustration of the types of links (user links and inter-tile links) and nodes (tiles and users) in a PWE.

The symmetry remark can be used for defining a common format for inputs and outputs in Section 9.2.2. It will also be called upon later on, to ensure that communication channels created by tuning HyperSurfaces are bidirectional.

EM functions f_h are a linear map of $\widetilde{E_{in}} \rightarrow \widetilde{E_{out}}$ [89]:

$$f_h \left(c \cdot \overrightarrow{E_{in}} + \sum_{\forall k} c_k \cdot \widetilde{E_{in}^k} \right) = c \cdot f_h \left(\overrightarrow{E_{in}} \right) + \sum_{\forall k} c_k \cdot f_h \left(\widetilde{E_{in}^k} \right) \quad (9.4)$$

where k is any index, and c, $c_k \in \mathcal{R}$.

The linearity property, in conjunction with the symmetry property will be promptly employed to reform the input/output format of f_h, without loss of generality.

9.2.2 Specialized Modeling of Function Inputs/Outputs

In communication scenarios, considering function input/outputs at the level of the EM field may not be practical. Instead, considering the signal source location and characteristics that yields the $\widetilde{E_{in}}$ can be more useful. To this end, we define the following input formats:

Planar wave. This case corresponds to a wave with a planar or almost planar wavefront, as shown in Fig. 9.3.

Planar waves can approximate:

■ waves impinging on a tile from a distant antenna at its far region,

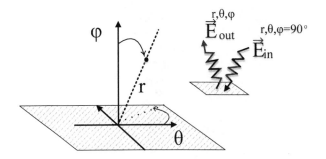

Figure 9.2 The tile coordinate system for describing its inputs and outputs. The origin is at the tile center.

■ waves that have been collimated at a preceding tile (*e.g.*, h_1 in Fig. 9.3), by applying the corresponding EM function.

Focusing on the second case, we will treat the collimation output as the source of the planar wave. Employing the coordinate system of the tile receiving this wave (h_2 in Fig. 9.3) and due to the planarity assumption, a single-frequency (ω) wave of this class can be simply described by:

■ its direction, $\overline{D}$: $\{r = \emptyset, \theta, \phi\}$ (using the $\emptyset$ notation to denote irrelevance from the r-dimension),

■ the total carried power, **P**, that impinges upon the surface of the tile (the summation of the Poynting vector norm over any bounded wavefront),

■ the normalized Jones vector, $\overrightarrow{\mathbf{J}}$ [41], describing the wave polarization at tile h_2,

■ the wavefront phase Φ at tile h_2.

Since the field $\widetilde{E_{in}}$ can be reconstructed from the aforementioned quantities, we proceed to replace it with the input parameter set for plane waves, $\widetilde{\mathbb{I}_{PW}}$:

$$\widetilde{E_{in}} \mapsto \widetilde{\mathbb{I}_{PW}} : \left\langle \omega, \overline{D}, \mathbf{P}, \overrightarrow{\mathbf{J}}, \Phi \right\rangle \tag{9.5}$$

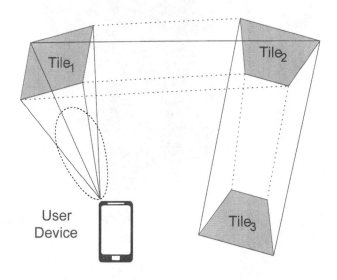

Figure 9.3 Illustration of the propagation wavefront between users and tiles.

Likewise, a planar output field $\widetilde{E_{out}}$ *defined over the surface of* h_2 can be alternatively expressed by the output parameter set for plane waves, $\widetilde{\mathbb{O}_{PW}}$:

$$\widetilde{E_{out}} \mapsto \widetilde{\mathbb{O}_{PW}} : \left\langle \omega', \overline{D}', \mathbf{P}', \overrightarrow{\mathbf{J}}', \Phi' \right\rangle \tag{9.6}$$

where the ($'$) notation implies generally different values than relation (9.5).

Focal wave. This case represents any generally non-collimated radiation from a mobile device (cf. Tx in Fig. 9.3), with its energy dissipating over an ever-growing sphere. In this case, the EM field impinging upon a tile also depends on the characteristics of the antenna device (radiation pattern and orientation). Assuming that this information is known and constant, and using a similar approach as above an input field at a tile, generated by such a source can be replaced as:

$$\widetilde{E_{in}} \mapsto \widetilde{\mathbb{I}_{FW}} : \left\langle \omega, \overline{D}, \mathbf{P}, \overrightarrow{\mathbf{J}}, \Phi \right\rangle \tag{9.7}$$

where $\overline{D} : \{r, \theta, \phi\}$, and $\overrightarrow{\mathbf{J}}$, Φ potentially vary over the tile surface. (I.e., $\overrightarrow{\mathbf{J}}(r, \theta)$, $\Phi(r, \theta)$). In the case where the focal wave is created (output) by a tile with the application of a proper function, the output field is similarly expressed as:

$$\widetilde{E_{out}} \mapsto \widetilde{\mathbb{O}_{FW}} : \left\langle \omega', \overline{D}', \mathbf{P}', \overrightarrow{\mathbf{J}}', \Phi' \right\rangle \tag{9.8}$$

The notations (9.5), (9.6), (9.7) and (9.8) are compatible with the Symmetry property, since $\mathbb{O}_{PW}$ and $\mathbb{O}_{FW}$ are also valid as inputs of any function f_h, while $\mathbb{I}_{PW}$ and $\mathbb{I}_{FW}$ are also valid as outputs of any function f_h at tile h.

Finally, the linear map property of relation (9.4) is rewritten as:

$$f_h\left(\{\mathbb{I}_1, \mathbb{I}_2, \mathbb{I}_3, \dots\}\right) = \{f_h\left(\mathbb{I}_1\right), f_h\left(\mathbb{I}_2\right), f_h\left(\mathbb{I}_3\right), \dots\} \qquad (9.9)$$

meaning that when multiple inputs are passed to a tile function, the total output is produced by applying the function to each input separately.

9.2.3 Modeling Core HyperSurface Functions

We proceed to study specialized HyperSurface functions, which are of practical value to the studied programmable wireless environments. Pure functions that can act as building blocks will be studied first, followed by a model for combining them into more complex ones.

Absorb f_h^{ABS}. Plane wave absorption has constituted one the most prominent showcases of metasurfaces [89]. In the studied programmable environments, absorbing unwanted reflections is important for interference minimization, as well as enforcing determinism over EM propagation. The ideal absorption function for an intended plane wave input is expressed as:

$$f_h^{\text{ABS}}\left(\mathbb{I}_{PW}\right) \to \emptyset \qquad (9.10)$$

where the empty set denotes no output wave. Full absorption is attained by matching the surface impedance of the tile to the incoming wave. For a planar input, this means that the surface impedance is constant across the tile, resulting into zero phase gradient and normal (specular) reflective behavior. This remark facilitates the modeling of unintended inputs as follows:

$$f_h^{\text{ABS}}\left(\widetilde{\mathbb{I}_{PW}} : \left\langle \omega, \overline{D}, \mathbf{P}, \vec{\mathbf{J}}, \Phi \right\rangle, \mathbb{I}_{PW}\right) \to$$
$$\left\langle \omega, \text{SPEC}\left(\overline{D}\right), g \cdot \mathbf{P}, \vec{\mathbf{J}}, \Phi \right\rangle \quad (9.11)$$

where $g\left(\mathbb{I}_{PW}, \widetilde{\mathbb{I}_{PW}}\right) \in \mathcal{R} < 1$ is a metric of similarity between $\mathbb{I}_{PW}$ and $\widetilde{\mathbb{I}_{PW}}$, defined by the physical structure of the HyperSurface and incorporating any constant material losses. The specular reflection of $\overline{D}$ is calculated as:

$$\text{SPEC}\left(\overline{D}\right) = \overline{D} - 2\left(\overline{D} \cdot \vec{n}\right)\vec{n} \qquad (9.12)$$

where $\overline{D}$ is directed and $\vec{n}$ is the unit normal of the tile surface.

It is noted that relation (9.11) can be employed to achieve an intended: i) partial attenuation of a wave, and ii) frequency selective absorption (filtering) [89].

Steer f_h^{STR}. Steering plain waves from an incoming direction to another is achieved by enforcing a gradient surface impedance that corresponds to the required reflection index [89]. This physical phenomenon can be described by considering a virtual surface normal, $\vec{n}'$, supplied as an input parameter of steering, that corresponds to a required reflection direction via relation (9.12). We proceed to define f_h^{STR} as follows:

$$f_h^{\text{STR}} \left(\widetilde{\mathbb{I}_{PW}} : \left\langle \omega, \overline{D}, \mathbf{P}, \vec{\mathbf{J}}, \Phi \right\rangle, \vec{n}', \mathbb{I}_{PW} \right) \to$$

$$\left\langle \omega, \overline{D} - 2 \left(\overline{D} \cdot \vec{n}' \right) \vec{n}', g \cdot \mathbf{P}, \vec{\mathbf{J}}, \Phi \right\rangle \quad (9.13)$$

where $g \left(\mathbb{I}_{PW}, \widetilde{\mathbb{I}_{PW}} \right) \in \mathcal{R} < 1$ in defined as in relation (9.11), noting that its specific expression generally differs from that of relation (9.11). We remark that the variant surface normal approach allows for a uniform expression (9.13), covering both the intended and unintended inputs.

Collimate f_h^{COL}. Collimation is the action of transforming a nonplanar input wave to a planar output [89]. In essence, the HyperSurface is configured for a virtual surface normal that varies across the tile surface, matching the local direction of arrival of the input. In the present scope, collimation will be studied for focal wave input as follows:

$$f_h^{\text{COL}} \left(\mathbb{I}_{FW}, \overline{D}' \right) \to \mathbb{O}_{PW} : \left\langle \omega, \overline{D}', g \cdot \mathbf{P}, \vec{\mathbf{J}}, \Phi \right\rangle \quad (9.14)$$

where the focal wave characteristics and the intended reflection direction are provided as inputs. Relation (9.14) refers to intended inputs and outputs. Obtaining the output to an unintended input can be based on the surface normal across the tile. Each sub-area over which the surface normal can be considered constant interacts with a part of the wavefront that can be considered planar, resulting into a reflection calculated via relation (9.12). Thus, the outcome to unintended input is a set of planar outputs, each with its own power, polarity and phase:

$$f_h^{\text{COL}} \left(\widetilde{\mathbb{I}_{FW}}, \mathbb{I}_{FW}, \overline{D}' \right) \to \{ \mathbb{O}_{PW} \} \quad (9.15)$$

In the context of the studied programmable environments, collimation is intended to be used mainly at the first and last hop of the propagation

from one device to another [95]. The first application (at a tile) transforms the waves emitted from a device to a planar form (see Fig. 9.3).
Tile-to-tile propagation is then performed for planar inputs. At the
final tile before reaching the receiver, the planar wave is focused to the
intended spot. This focusing is essentially collimation in the reverse,
where a planar wave is converted to focal output. The tile configuration
for focusing remains the same as in collimation, due to the symmetry
in Remark 9.2.1.

Polarize m_h^{POL}. The physics of polarization control can be
described qualitatively by assuming equivalent circuits of meta-
atoms [89]. In essence, each meta-atom can be viewed as a circuit
with a input and output antennas, as well as cross-couplings among
meta-atoms. A wave enters via the input antennas, undergoes some
alteration via the circuit and exits via output antennas, subject to the
connections performed by tuning the active elements. Polarization is
thus a shift in the Jones vector, attained by the appropriate choice of
output antennas. Pure polarization control does not affect other wave
parameters. As such, we define the polarization control not as a stand-
alone function, but rather as a modifier applied to the output of a
preceding function:

$$m_h^{\text{POL}}\left(\mathbb{O}_{PW} : \left\langle \omega, \overline{D}, \mathbf{P}, \overrightarrow{\mathbf{J}}, \Phi \right\rangle, \Delta\overrightarrow{\mathbf{J}}\right) \rightarrow$$
$$\mathbb{O}'_{PW} : \left\langle \omega, \overline{D}, \mathbf{P}, \overrightarrow{\mathbf{J}} + \Delta\overrightarrow{\mathbf{J}}, \Phi \right\rangle \quad (9.16)$$

For unintended outputs, the intended shift $\Delta\overrightarrow{\mathbf{J}}$ is not necessarily
attained, which can be expressed as:

$$m_h^{\text{POL}}\left(\widetilde{\mathbb{O}_{PW}}, \mathbb{O}_{PW}, \Delta\overrightarrow{\mathbf{J}}\right) \rightarrow$$
$$\widetilde{\mathbb{O}'_{PW}} : \left\langle \omega, \overline{D}, \mathbf{P}, \overrightarrow{\mathbf{J}} + \Delta\overrightarrow{\mathbf{J}}', \Phi \right\rangle \quad (9.17)$$

where $\Delta\overrightarrow{\mathbf{J}}' \neq \Delta\overrightarrow{\mathbf{J}}$ in general. Notice that power loss concerns are
delegated to the preceding pure function.

Phase Alteration m_h^{PHA}. Phase alteration follows the same principle as the polarization. Qualitatively, the modification of the wave
phase is accomplished within the equivalent circuit, via an inductive
or capacitive element [89]. Once again, this functionality is defined

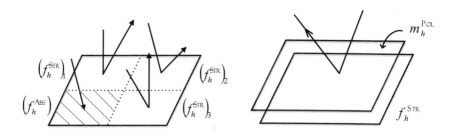

Figure 9.4 Function combination models: Surface Division (left) and Meta-atom Merge (right).

as a modifier:

$$m_h^{\text{PHA}}\left(\mathbb{O}_{PW} : \left\langle \omega, \overline{D}, \mathbf{P}, \vec{\mathbf{J}}, \Phi \right\rangle, \Delta\Phi\right) \rightarrow$$

$$\mathbb{O}'_{PW} : \left\langle \omega, \overline{D}, \mathbf{P}, \vec{\mathbf{J}}, \Phi + \Delta\Phi \right\rangle \quad (9.18)$$

For unintended outputs, the definition is altered similarly to relation (9.17).

Combination model. The pure functions studied above can be combined to describe a more complex functionality. Complex functions may be an operational requirement (e.g., steer and polarize at the same time) or be imposed by physical imperfections of the metasurface (e.g., being unable to steer without altering the polarization). We proceed to present a common model to describe both cases.

Surface division. This combination approach assigns different functions to different sub-areas of the same tile [196]. The principle of operation is shown in Fig. 9.4 (left inset), where an impinging plane undergoes splitting into three separate directions and partial attenuation. The power distribution of any output waves depends on the area allocated to each sub-function. For instance, an ideal N-way split of an input $\mathbb{I}_{PW}$ towards directions with custom normals $\vec{n}_i'$, $i = 1 \ldots N$ can be expressed as:

$$f_h^{\text{SPL}}\left(\mathbb{I}_{PW}, \{\vec{n}_i'\}\right) \rightarrow \{\mathbb{O}_{PW}\},$$

$$\{\mathbb{O}_{PW}\} : \left\{\left\langle \omega, \overline{D}-2(\overline{D}\cdot\vec{n}_i')\vec{n}_i', \tfrac{1}{N}\cdot\mathbf{P}, \vec{\mathbf{J}}, \Phi\right\rangle \big| i = 1 \ldots N\right\} \quad (9.19)$$

where $(\mathbf{P}/N)$ is the power of each output.

Table 9.2 Combinable tile functions and methods.

		Merge with				
		f_h^{ABS}	f_h^{STR}	f_h^{COL}	m_h^{POL}	m_h^{PHA}
Function	f_h^{ABS}	SD	SD	SD	MM*	MM*
	f_h^{STR}	SD	SD	SD	MM	MM
	f_h^{COL}	SD	SD	SD	MM	MM
	m_h^{POL}	MM*	MM	MM	-	MM
	m_h^{PHA}	MM*	MM	MM	MM	-

SD: Surface Division, **MM:** Meta-atom Merge,
$\sim$: Possible but potentially unintended.
*: Defined only when f_h^{ABS} produces output (partial absorb).

Meta-atom merge. A common practice in metasurface analysis is to merge meta-atoms to create more complex basic structures, called supercells [63]. Essentially, using the equivalent circuit paradigm, merging meta-atoms creates circuits with more degrees of tunability. This enables the combination of functions and modifiers, e.g., for steering and polarizing or steering and phase altering at the same time, over the same surface (cf. Fig. 9.4, right inset), denoted as:

$$m_h^{\text{POL}}\left(f_h^{\text{STR}}\right), \, m_h^{\text{PHA}}\left(f_h^{\text{STR}}\right), \text{ or } m_h^{\text{POL}}\left(m_h^{\text{PHA}}\left(f_h^{\text{STR}}\right)\right) \qquad (9.20)$$

Additionally, it is possible to apply a meta-atom merge-derived function only to a sub-area of a tile, thus combining it with surface division.

A summary of combinable functions and combination approaches is given in Table 9.2. Notably: i) the combination patterns are symmetric, ii) combining collimation with any other non-modifier function is possible but potentially unintended, as described in the context of relation (9.15), iii) combining a modifier with absorption makes sense only when the absorption is partial (i.e., there is an output to apply the modifier upon), and iv) combining several modifiers of the same type to the same tile is the same as applying it once with the total modification. As such, these combinations are trivial. *Finally, it is noted that any number of functions (i.e., more than two) can be merged via the surface division model.*

It is worth noting that merging functions is not without impact on the efficiency of the HyperSurface. Any metasurface requires a minimum amount of meta-atoms to yield a consistent behavior (i.e., with near-unitary efficiency) [196]. Both surface division and meta-atom merge naturally limit the meta-atom numbers available for a given functionality. As such, combining functions will generally amplify discretization, boundary and other errors [43]. This can lead to reduced efficiency, which can be expressed as considerably attenuated intended outputs, as well as the appearance of unintended, parasitic outputs, $\mathbb{O}_\mathbb{P}$, even for intended inputs:

$$f_h \left(\mathbb{I}, \widetilde{\mathbb{I}} \right) \rightarrow f_h \left(\mathbb{I} \right) + f_h \left(\widetilde{\mathbb{I}} \right) + \mathbb{O}_\mathbb{P} \qquad (9.21)$$

where $f_h \left(\mathbb{I} \right)$ is the intended output and $f_h \left(\widetilde{\mathbb{I}} \right)$ is the well-defined output to unintended input. Thus:

Combining functionalities over a single tile generally reduces the efficiency of the overall function.

The effects of Remark 9.2.3 can be quantified only per specific, physical HyperSurface design. Nonetheless, we will employ this remark in ensuing sections and setup a policy of minimizing function combinations in programmable wireless environments.

The generic relation (9.21), in conjunction with the preceding modeling defines the information that a HyperSurface manufacturer should measure and provide, to facilitate the use in programmable environments. This information, collectively referred to as *EM profile* contains the following information:

◼ The supported function types and allowed combinations, as a subset of the entries of Table 9.2.

◼ The intended and parasitic outputs to intended inputs.

◼ The unintended and parasitic outputs to unintended inputs.

The EM profile can be obtained, e.g., by measuring the scattering pattern in controlled conditions, for the complete array of supported functions and intended inputs. Nonetheless, an exhaustive measurement for any unintended input may be prohibitive. In this case, the profile may provide a calculation model for outputs, such as the $g \left(\mathbb{I}_{PW}, \widetilde{\mathbb{I}_{PW}} \right)$ metric employed in Section 9.2.3.

For the remainder of this study, we will consider the EM profile as a given, provided by the tile manufacturer.

9.2.4 A Graph Model for Simulating and Optimizing Programmable Wireless Environments

Propagation within a 3D space comprises Line-of-sight (LOS) and Non-LOS (NLOS) components. Naturally, PWEs control the NLOS component only, without affecting the LOS [92]. Therefore, the following model will focus on the NLOS case.

Consider two tiles, h and h', in a 3D space. We will consider these tiles as *connectable* if there exist any input $\widetilde{\mathbb{I}}$ and functions f_h, $f_{h'}$, such that:

$$f_{h'}\left(f_h\left(\widetilde{\mathbb{I}}\right)\right) \to \emptyset \iff f_{h'}: f_{h'}^{ABS}, \forall f_h\left(\widetilde{\mathbb{I}}\right) \neq \emptyset \qquad (9.22)$$

In other words, inter-tile connectivity means that one tile can redirect impinging EM energy to another. It is implied that the redirected power is significant, i.e., it surpasses a practical threshold defined by the application scenario. Thus, connectability may be precluded due to physical obstacles between tiles or by the lack of supported tile functions to redirect significant energy to one another.

Additionally, we consider a set of user devices, $\mathcal{U}$, in the same space. User $u \in \mathcal{U}$ will be considered *connected* to tile h if there exists any LOS input $\widetilde{\mathbb{I}}$ and a function f_h such that:

$$f_h\left(\widetilde{\mathbb{I}}\right) \to \emptyset \iff f_h: f_h^{ABS}, \forall\widetilde{\mathbb{I}} \neq \emptyset \qquad (9.23)$$

i.e., getting zero output from a tile connected to a user is only possible if a full absorption functionality is employed.

Based on relations (9.22) and (9.23), we define the notion of:

■ the set of inter-tile links, $\mathcal{L}_h$, where each contained link $l_{h,h'} \in \mathcal{L}_h$ denotes tile *connectivity potential* by relation (9.22), and

■ the set of user-tile links, $\mathcal{L}_u$, where each contained link $l_{u,h} \in \mathcal{L}_u$ denotes *connection* by relation (9.23).

Following the symmetry Remark 9.2.1, all links in $\mathcal{L}_h$ and $\mathcal{L}_u$ are bidirectional and symmetric. Additionally all links will be considered to have an associated label, DELAY (l), defined as the wave propagation delay, inclusive of any delay within the receiving end. Finally, last-tile-*to*-receiving-user links may be labeled as Tx $(l_{u,h}) \in \mathcal{U}$, to designate the transmitting user that may employ them. Likewise, transmitting-user-*to*-first-tile links may be labeled as Rx $(l_{u,h}) \in \mathcal{U}$, to designate the receiving user that must be reached via them. The labeling is intended to capture the Multiple-Input-Multiple-Output (MIMO) potential of the user devices at a high level.

Based on the above definitions, we proceed to define the graph $\mathcal{G}\left\langle\{\mathcal{H}, \mathcal{U}\}, \{\mathcal{L}_h, \mathcal{L}_u\}\right\rangle$, as well as subgraphs of the form $\mathcal{G}\left(\{f_h\}\right)$. The latter represent environments that have been configured by applying specific functions to tiles, thus instantiating some of the allowed links $\mathcal{L}_h$. We consider an EM flow, $\langle u_{\mathrm{TX}}, u_{\mathrm{RX}}\rangle$, from a transmitter u_{TX} to a receiver u_{RX}, following a path via tiles within $\mathcal{G}\left(\{f_h\}\right)$, defined as an ordered selection of links without repetitions:

$$\overrightarrow{p}_{\langle u_{\mathrm{TX}} u_{\mathrm{RX}}\rangle} = \left\{l_{u_{\mathrm{TX}} h_1}, l_{h_1 h_2}, l_{h_2 h_3}, \ldots, l_{h_N u_{\mathrm{RX}}}\right\}. \tag{9.24}$$

The above formulation with non-repeating links in paths is also posed in compliance with Remark 9.2.3, which dictates the avoidance of function combinations. To understand this claim, consider a counter-example with a repetitive path. This path must have the form $\left\{\ldots l_{h_k h_i}, l_{h_i h_{i+1}}, \ldots l_{h_l h_i}, l_{h_i h_{i+1}}, \ldots\right\}$ where $k \neq l$. Due to the symmetry Remark 9.2.1, a reverse input via the repeated link $l_{h_i h_{i+1}}$ must exit from both $l_{h_k h_i}$ and $l_{h_l h_i}$. This requires a splitting function at tile h_i, which is a combined function, as described in the context of relation (9.19).

Any input $\widetilde{\mathbb{I}}$ at tile h_1 is transformed over the path as:

$$\widetilde{\mathbb{O}}\left(\widetilde{\mathbb{I}}, \overrightarrow{p}_{\langle u_{\mathrm{TX}} u_{\mathrm{RX}}\rangle}\right) \leftarrow f_{h_N} \circ \ldots f_{h_3} \circ f_{h_2} \circ f_{h_1} \circ \widetilde{\mathbb{I}} \tag{9.25}$$

where $f_{h_j} \circ f_{h_i}$ implies passing only the output of f_{h_i} that impinges tile h_j to f_{h_j}, and ignoring any other outputs of f_{h_i}. Without loss of generality, relation (9.25) omits the propagation over the first and last link, which are subject to antenna characteristics and standard (non-programmable) propagation. The end-to-end delay of a path can be calculated in a trivial manner as the sum of all link delays:

$$\mathrm{DELAY}\left(\overrightarrow{p}_{\langle u_{\mathrm{TX}} u_{\mathrm{RX}}\rangle}\right) = \sum_{\forall l \in \overrightarrow{p}} \mathrm{DELAY}\left(l\right) \tag{9.26}$$

We proceed to consider all paths that reach the receiver u_{RX}, from any transmitter $(*)$ (noting the possible multiplicity of paths for the same $\langle u_{\mathrm{TX}}, u_{\mathrm{RX}}\rangle$ pair). The total received output is then:

$$\widetilde{\mathbb{O}}_{u_{\mathrm{RX}}}^{\mathrm{TOTAL}} = \sum_{\forall \overrightarrow{p} \,:\, \overrightarrow{p}\langle *, u_{\mathrm{RX}}\rangle} \widetilde{\mathbb{O}}\left(\widetilde{\mathbb{I}}, \overrightarrow{p}\right) \tag{9.27}$$

Relations (9.26), (9.27) contain all the required information for deriving the communication quality of a pair $\langle u_{\mathrm{TX}}, u_{\mathrm{RX}}\rangle$. Moreover, due

Algorithm 1 Simulated NLOS propagation model for PWEs.

1: **procedure**   PATHS$_{\text{RX}}$:   NLOSPROP$(\mathcal{G}(\{f_h\}), u_{\text{RX}},$
$\left\{\left\langle u_{\text{TX}}, \left\{\widetilde{\mathbb{I}}_h, h : l_{u,h} \in \{\mathcal{L}_u : u = u_{\text{TX}}\}\right\}\right\rangle\right\})$

2:      rays $\leftarrow \{\}$, PATHS$_{\text{RX}} \leftarrow \{\}$;

3:      **for** $l_{u,h} \in \{\mathcal{L}_u : u \in \{u_{\text{TX}}\}\}$ //for all transmitters.

4:        $r \leftarrow \left\langle \overrightarrow{p}_r = \{l_{u,h}\}, \widetilde{\mathbb{I}}_r = \widetilde{\mathbb{I}}_h \right\rangle$;

5:        rays $\leftarrow$ rays $+ r$;

6:      **end for**

7:      **while** rays $\neq \emptyset$

8:        $r \leftarrow$ rays.pop();   //remove and return last entry.

9:        **for** $o \in \widetilde{\mathbb{O}}\left(r : \widetilde{\mathbb{I}}_r, r : \overrightarrow{p}_r\right)$ //cf. rel. (9.25).

10:          $h^* \leftarrow$ TILEREACHED (o);   //cf. rel. (9.22).

11:          $u^* \leftarrow$ USERREACHED (o);   //cf. rel. (9.23).

12:          $h \leftarrow$ LASTTILE $(r : \overrightarrow{p}_r)$;

13:          **if** $(u^* = u_{\text{RX}})$

14:            $r^* \leftarrow \left\langle r : \overrightarrow{p}_r + l_{h,u^*}, r : \widetilde{\mathbb{I}}_r \right\rangle$;

15:            PATHS$_{\text{RX}} \leftarrow$ PATHS$_{\text{RX}} + (r : \overrightarrow{p}_r + l_{h,u^*})$;

16:          **else if** $(h^* \neq \emptyset)$

17:            $r^* \leftarrow \left\langle r : \overrightarrow{p}_r + l_{h,h^*}, r : \widetilde{\mathbb{I}}_r \right\rangle$;

18:            rays $\leftarrow$ rays $+ r^*$;

19:          **end if**

20:        **end for**

21:      **end while**

22: **end procedure**

to the symmetry Remark 9.2.1, the derivations of relations (9.26), (9.27) are identical for the reverse path, $\langle u_{\text{RX}}, u_{\text{TX}}\rangle$. Finally, the labeling $\text{Tx}(l_{h_N u_{\text{RX}}})$ of the receiver's links can be employed to classify outputs as useful or interfering based on the MIMO configuration.

The paths $\overrightarrow{p}_{\langle u_{\text{TX}} u_{\text{RX}}\rangle}$ are directly derived from the deployed tile functions and the corresponding graph $\mathcal{G}(\{f_h\})$. The simulated NLOS propagation procedure is modeled as *NLOSprop* (Algorithm 1). The process receives as inputs: i) the configured tiles, ii) a receiver, and iii) all transmitting users and their inputs ($\widetilde{\mathbb{I}}$ for each link connecting the transmitter to a tile). The process produces the paths leading from any transmitter to the receiver. Internally, a tuple denoted as *ray* is used for holding a path (cf. rel. (9.24)) and its input. A stack of *rays* is initialized per transmitter link (lines $3-6$). The model then continuously updates each ray, accounting for tile interactions (line 9). It is noted that each interaction may yield more than one output. Once a path reaches the intended user, it is added to the model outputs (lines $13-15$) and the corresponding ray is discarded. If a tile is reached, a copy of the ray with an updated path is added to the ray set for further processing (lines $16-19$). The model accounts for attenuated rays, by using the connectivity definitions (9.22), (9.23)–weak rays are treated as null outputs. Furthermore, notice that rays escaping the considered space are discarded as expected (lines $10, 11$ will yield null outputs). Finally, notice that the model can also provide the paths to a set of receivers, $\{u_{\mathrm{RX}}\}$, by modifying the condition in line 13 as $u^* \in \{u_{\mathrm{RX}}\}$.

Programmable wireless environments seek to optimize the communication for any set $\langle \{u_{\mathrm{RX}}\}, \{u_{\mathrm{TX}}\} \rangle$, by deploying the corresponding, performance optimizing tile functions. This can be generally formulated as:

$$\{f_h\}^{OPT} \leftarrow \underset{\{f_h\}}{\arg \mathrm{opt}} (\mathrm{OBJECTIVE}(\mathrm{LOSPROP}(\langle \{u_{\mathrm{RX}}\}, \{u_{\mathrm{TX}}\} \rangle)$$

$$+ \mathrm{NLOSPROP}\left(\mathcal{G}\left(\{f_h\}\right), \langle \{u_{\mathrm{RX}}\}, \{u_{\mathrm{TX}}\} \rangle \right))) \quad (9.28)$$

where *objective* is a fitness function applied to the propagation outcome, and the inputs $\overline{\mathbb{I}}$ of *NLOSProp* are omitted for clarity. The *objective* may freely refer to:

■ 1 transmitter to 1 receiver (uni-cast), $\langle u_{\mathrm{TX}}, u_{\mathrm{RX}} \rangle$,

■ 1 transmitter to many receivers (multi-cast or broad-cast), $\langle u_{\mathrm{TX}}, \{u_{\mathrm{RX}}\} \rangle$,

■ many transmitters to many receivers, $\langle \{u_{\mathrm{TX}}\}, \{u_{\mathrm{RX}}\} \rangle$.

The latter two categories inherently incorporate resource sharing policies among communicating pairs.

It is noted that the formulation of relation (9.28) is generic and, thus, also covers cases that are not compliant to Remark 9.2.3 about tile functionality reuse.

9.2.5 Modeling Connectivity Objectives

We proceed to study specific objectives for core aspects of wireless system performance. We focus on objectives pertaining to a single receiver, since resource sharing policies are subjective. It is noted, however, that

Section 9.3 case-studies the integration of objectives and policies. Additionally, all *LOS* components will be omitted for clarity, since they are either not affected by PWEs or will be manipulated for control in the same manner as NLOS.

We proceed to define a path-centric formulation for various core objectives. In other words, we employ the correspondence $\{f_h\} \leftrightarrow \overrightarrow{p}$ between a set of deployed functions f_h and formed paths $\overrightarrow{p}$ interconnecting the various users within the PWE. The practicality of this choice will be explained in the remainder of the current subsection and in Section 9.3.

9.2.5.1 Power Transfer Maximization

We study the objective of maximizing the total power received by a specific user, emitted by a group of transmitters. This objective is practical for wireless power transfer [92], as well as for formulating subsequent objectives. Using a path-centric approach, relation (9.28) can be rewritten as:

$$\{\overrightarrow{p}\} \overset{OPT}{\underset{\{\overrightarrow{p}\}}{\leftarrow} \arg\max} \left(\sum_{\forall u^* \in \{u_{\mathrm{TX}}\}} \sum_{\forall i} \widetilde{\mathbb{O}} \left(\widetilde{\mathbb{I}}_{u^*,i}, \ \overrightarrow{p_i}_{\langle u^*, u_{\mathrm{RX}} \rangle} \right) \right) \qquad (9.29)$$

i.e., the maximization of the aggregate output of all paths leading to the studied receiver, u_{RX}, where i indexes multiple paths for the same transmitter-receiver pair. Since the objective is the maximization of power, only the **P** attribute of $\widetilde{\mathbb{O}}$ needs to be retained (cf. def. (9.6), (9.8)). Moreover, the output can be expressed as the product of all gain metrics g_h (defined in Section 9.2.3) over the tiles comprising each path, multiplied by the input power of the path:

$$\{\overrightarrow{p}\} \overset{OPT}{\underset{\{\overrightarrow{p}\}}{\leftarrow} \arg\max} \left(\sum_{\forall u^* \in \{u_{\mathrm{TX}}\}} \sum_{\forall i} \mathbf{P}_{u^*,i} \cdot \prod_{\forall h \in \overrightarrow{p_i}_{\langle u^*, u_{\mathrm{RX}} \rangle}} g_h \right) \qquad (9.30)$$

It is noted that the path-centric formulation allows for employing the $\prod g_h$ expression in relation (9.30). In other words, treating a path as input to the optimization, allows for calculating (and caching) its total effect on the transmitter power. An alternative f_h-centric formulation would not allow this simple expression, since the output of a tile–and, thus, its gain–depends on the input wave and the deployed tile function (e.g., cf. rel. (9.13)). Thus, a function-centric formulation would have to follow expression (9.25), where each gain would receive as input the preceding gains and be passed as argument to the next one, i.e., $g\left(g\left(g\left(\ldots\right)\right)\right)$.

9.2.5.2 QoS Optimization

QoS refers to optimizing some aspect of the communication channel between two users. Without loss of generality, we will focus on the maximization of the useful-signal-to-interference ratio $(\mathbf{P}^{TOT}/\mathbf{I})$ at the studied receiver, u_{RX}. The useful received signal is expressed as:

$$\mathbf{P}^{TOT} = \sum_{\forall u^* \in \{u_{\mathrm{TX}}\}} \sum_{\forall i} \mathbf{P}_{u^*,i} \cdot \prod_{\forall h \in \overrightarrow{p_i}_{\langle u^*, u_{\mathrm{RX}} \rangle} : \mathcal{P}'} g_h \qquad (9.31)$$

where

$$\mathcal{P}' : \{\overrightarrow{p_i}_{\langle u^*, u_{\mathrm{RX}} \rangle} \in \mathcal{P} | \mathrm{DELAY}\left(\overrightarrow{p_i}_{\langle u^*, u_{\mathrm{RX}} \rangle}\right) - \\ \min\left(\{\mathrm{DELAY}\left(\overrightarrow{p} \in \mathcal{P}\right)\}\right) < \mathrm{D_{TH}}\} \qquad (9.32)$$

$$\mathcal{P} : \{\overrightarrow{p_i}_{\langle u^*, u_{\mathrm{RX}} \rangle} | \mathrm{Tx}\left(\mathrm{LASTLINK}\left(\overrightarrow{p_i}_{\langle u^*, u_{\mathrm{RX}} \rangle}\right)\right) = u^*, \\ \mathrm{Rx}\left(\mathrm{FIRSTLINK}\left(\overrightarrow{p_i}_{\langle u^*, u_{\mathrm{RX}} \rangle}\right)\right) = u_{\mathrm{RX}}\} \qquad (9.33)$$

Restriction (9.33) (defining path set $\mathcal{P}$) states that a path carries a useful signal: i) if the transmission is intended for the studied receiver, and ii) is received by a link labeled to expect the transmitter emission. Therefore, it is implied that the MIMO capabilities of the user devices have also been configured. (This is accomplished in Section 9.3). Restriction (9.32) defines a subset of $\mathcal{P}$, based on the latency of each path. Accounting for constructive signal superposition, the selected paths carrying useful signals must have bounded latency (i.e., within a time window defined by the fastest path, plus an application-specific threshold value $\mathrm{D_{TH}}$).

The interference is expressed similarly to $\mathbf{P}^{TOT}$ but over all paths $\overline{\mathcal{P}'}$ leading to the studied receiver, but not found in $\mathcal{P}'$:

$$\mathbf{I} = \sum_{\forall u^* \in \{u_{\mathrm{TX}}\}} \sum_{\forall i} \mathbf{P}_{u^*,i} \cdot \prod_{\forall h \in \overrightarrow{p_i}_{\langle u^*, u_{\mathrm{RX}} \rangle} : \overline{\mathcal{P}'}} g_h \qquad (9.34)$$

Finally, the QoS optimization objective is expressed as:

$$\{\overrightarrow{p}\} \overset{OPT}{\leftarrow} \underset{\{\overrightarrow{p}\}}{\arg\max} \left(\frac{\mathbf{P}^{TOT}}{\mathbf{I}}\right) \qquad (9.35)$$

9.2.5.3 Eavesdropping Mitigation

Security concerns pertaining to eavesdropping can be taken into account and be combined with other objectives [95]. Eavesdropping mitigation is achieved by ensuring that the employed communication paths avoid all but the intended user. To this end, consider a 3D surface, $S(u)$, around a user u. $S(u)$ can exemplary be a sphere centered at the user's position. We proceed to discard the paths produced by Algorithm 1 that geometrically intersect with the $S(u)$ (i.e., a check for intersection between any link in the path and the surface). Avoiding any user but the intended receiver can then be expressed as:

$$\overrightarrow{p_i}_{\langle u^*, u_{\mathrm{RX}} \rangle} | \overrightarrow{p_i} \cap S(u) \to \emptyset, \ \forall u \neq u_{\mathrm{RX}}, u \in \mathcal{U} \tag{9.36}$$

Restriction (9.36) can be further customized, e.g., to avoid a targeted set of users only, to discard paths prone to eavesdropping based on the intersection with some convex hull surface covering many users or even to quantify the eavesdropping risk as a scalar value, such as the minimum distance between a path and a user device.

The expression of a security concern as a restriction over paths, allows for natural combination between security and QoS or power transfer objectives. Restriction (9.36) can simply filter out paths before $\mathcal{P}$ (restriction (9.33)) or after $\mathcal{P}'$ (restriction (9.32)). In the former case security is prioritized over connectivity, while in the latter case connectivity comes first.

The described eavesdropping mitigation approach naturally limits the interference caused to unintended recipients. Thus, the eavesdropping mitigation is not antagonistic to other QoS objectives.

9.2.5.4 Doppler Effect Mitigation

The geometric approach in path restrictions can be extended to mitigate Doppler effects. Frequency shifts owed to Doppler effects are especially important in mm-wave communications (and higher frequencies), where even pedestrian movement rapidly deteriorates the reception quality [57]. To this end, PWE can strive to keep the last link of communication paths perpendicular to the trajectory of the user, $\overrightarrow{m}_{u_{\mathrm{RX}}}$, as follows:

$$\overrightarrow{p_i}_{\langle u^*, u_{\mathrm{RX}} \rangle} | \mathrm{LastLink}(\overrightarrow{p_i}) \perp \overrightarrow{m}_{u_{\mathrm{RX}}} \tag{9.37}$$

Restriction (9.37) can alternatively be relaxed to a scalar metric of perpendicularity, e.g., as the inner product of the unary vector across

LASTLINK $(\overrightarrow{p_i})$ and the unary derivative of $\overrightarrow{m}_{u_{\text{RX}}}$ at the user position. Finally, restriction (9.37) can be combined with all preceding objectives in a manner identical to the eavesdropping case.

9.2.5.5 User Blocking

Blocking a user from gaining access to another device, such as an access point, can be facilitated by configuring a PWE to absorb its NLOS emissions. In this manner, potential security risks can be mitigated at the physical layer, before expending resources for blocking them at a higher level (e.g., MAC slots, software authorization steps) [92, 93, 95]. The mathematical formulation follows naturally from relation (9.30), by replacing max with min within the objective. It is implied that information on user authorization is passed to the PWE configuration as an input.

Finally, it is noted that this objective needs *not* be combined with any other, given that it intends to fully block the physical connectivity of a user, rather than assign resources to him.

9.3 A K-PATHS APPROACH FOR MULTI-USER MULTI-OBJECTIVE ENVIRONMENT CONFIGURATION

The preceding section formulated the PWE configuration objectives, using graph paths as inputs. We proceed to define the KpConfig heuristic, which configures a PWE for serving a set of user objectives (Algorithm 2). KpConfig receives the following input parameters:

■ The PWE graph $\mathcal{G}$ comprising the sub-graphs of connected users, $\mathcal{G}\langle\mathcal{U}, \mathcal{L}_u\rangle$, and connectable tiles, $\mathcal{G}\langle\mathcal{H}, \mathcal{L}_h\rangle$. We consider the latter as static and, therefore, pre-processable. Particularly, we assume that a custom number of node-disjoint shortest paths can be pre-calculated and cached, for each tile pair. The link weight for such calculations is the link *Delay*, and the ensuing paths are considered to be filtered based on their total steering gain (cf. the $\prod g_h$ expression in relation (9.30)) and a custom acceptable threshold. Subsequent calls to well-known path finding algorithms (e.g., *ShortestPath, KShortestPaths* [28]) are considered to be executed on top of this cache.

■ The set of communicating user pairs and their objectives. Multicast groups are expressed as multiple pairs with the same transmitter. The objective *obj* is a set of binary flags, each denoting a type from Section 9.2.5. We assume that symmetric pairs have

---

Algorithm 2 The K-Paths approach for configuring PWEs.

1: **procedure** $KpConfig(\mathcal{G},\{\langle u_{\text{TX}}, u_{\text{RX}}, \text{OBJ}\rangle\},$
   $\left\{\left\langle u_{\text{TX}}, \left\{\widetilde{\mathbb{I}}_h, h : l_{u,h} \in \{\mathcal{L}_u : u = u_{\text{TX}}\}\right\}\right\rangle\right\})$

2:      *blocked***pairs**,   *sorted***pairs**, N_e, K_e $\leftarrow MDFPolicy($
   $\mathcal{G},\{\langle u_{\text{TX}}, u_{\text{RX}}, \text{OBJ}\rangle\});$

3:      $pathsrem \leftarrow 0;$

4:      **for** $e : \langle u_{\text{TX}}, u_{\text{RX}}, \text{OBJ}\rangle$ in *sorted***pairs**

5:        $paths \leftarrow \{\emptyset\};$

6:        **for** $i = 1 : 1 : K_e$

7:          $\mathcal{L}_a \leftarrow \text{FILTERLINKSBYOBJ}(\mathcal{G}, \text{OBJ});$

8:          $\overrightarrow{p} \leftarrow \text{FINDCOMPLEXPATH}(\mathcal{G}, u_{\text{TX}}, u_{\text{RX}}, \mathcal{L}_a, paths);$

9:          **if** $\overrightarrow{p} = \emptyset$

10:            **break;**

11:          **end if**

12:          $paths \leftarrow paths + \overrightarrow{p};$

13:        **end for**

14:        **if** $\|paths\| = 0$

15:          **continue;** *//pair is disconnected.*

16:        **end**

17:        $maxpaths \leftarrow \min\{N_e + pathsrem, \|paths\|\};$

18:        $paths \leftarrow \text{FILTERPATHSBYOBJ}(paths, maxpaths, \text{OBJ});$

19:        $pathsrem \leftarrow pathsrem + \max\{maxpaths - \|paths\|, 0\};$

20:        $\text{DEPLOY}(\mathcal{G}, paths);$

21:      **end for**

22:      $\text{DEPLOYBLOCKS}(\mathcal{G}, blocked\textbf{pairs});$

23:      $f_h \leftarrow f_h^{ABS}, \forall h \in \mathcal{G} : f_h = \emptyset;$

24: **end procedure**

---

been filtered out of this parameter. Complimenting Remark 9.2.1, a pair $\langle u_{\text{TX}}, u_{\text{RX}}, \text{OBJ}\rangle$ is symmetric to $\langle u_{\text{RX}}, u_{\text{TX}}, \text{OBJ}\rangle$, highlighting that the objective must also be the same. Moreover, disconnected pairs (i.e., without any *ShortestPath* in $\mathcal{G}$) are also considered as filtered out.

■ The inputs $\widetilde{\mathbb{I}}_h$ and the affected tiles, for each transmitter.

The use of the link *Delay* as the link weight prioritizes shorter links to assemble a path. In practice, the output of a tile function may digress from the intended as the distance from the tile increases [89, 196]. In this aspect, shorter links may favor a more consistent behavior.

Additionally, the output of KpConfig is a set of paths per pair (which correspond to tile functions as described in Section 9.2.5). Thus, the Tx and Rx labels per user link are naturally produced by KpConfig, without further steps.

Since KpConfig seeks to serve multiple user pairs, a resource sharing policy needs to be employed in the general case. Thus, at line 2, KpConfig, calls upon the 'Most-Distant-First' (*MDFPolicy*) subroutine, which is an exemplary policy.

MDFPolicy, formulated as Algorithm 3, primarily returns a sorting of the user pairs (*sorted**pairs***) by descending priority order, as well as the number of paths to allocate per pair e, N_e. At lines $9 - 21$, *MDFPolicy* filters out any pairs with user access *block* objectives, since these will not require connecting paths. Remaining pairs are sorted by descending average delay, calculated over K_e shortest paths in graph $\mathcal{G}$, K_e being the minimum number of user links in the pair e. This simple heuristic intends to prioritize distant pairs, on the grounds of experiencing a lesser degree of propagation control. Finally, the tentative number of paths, N_e, to allocate pair is returned at line 24, as the minimum ratio of number of user links divided by the number of pairs this user belongs to.

KpConfig (Algorithm 2) then resumes its operation. In general, it comprises an exploratory evaluation of K_e paths per sorted pair (lines $6 - 13$), eventually keeping and deploying at most N_e subject to compliance with objectives (lines $17-20$). Left-over path allocations are redistributed to ensuing pairs via the *paths**rem*** variable (lines $3, 19$).

The exploration of K_e paths at lines $6 - 13$ is also a point for enforcing eavesdropping and Doppler effect mitigation objective types. At line 7, the *FilterLinksByObj* subroutine is called Algorithm 4, which returns the links $\mathcal{L}_a$ of $\mathcal{G}$ that are in conflict with the objectives. The use of `switch` without `break` statements implies that both objectives can be active for the same user pair.

At line 7 of KpConfig (Algorithm 2), the links $\mathcal{L}_a$ are passed to the *FindComplexPath* subroutine (Algorithm 5), which seeks to find a path in $\mathcal{G}$ that connects the pair while avoiding the links $\mathcal{L}_a$ and the links over the already deployed *paths*. Two subgraphs are created at lines 2 and 3 of Algorithm 5. The subgraph $\mathcal{G}'$, which removes the aforementioned links from the original graph $\mathcal{G}$, and $\mathcal{G}^*$ which also removes the already configured tiles (nodes) of $\mathcal{G}'$. *FindComplexPath* first tries to find a connecting path in $\mathcal{G}^*$ (line 4). If found, the path comprises unconfigured tiles only, thus being in compliance with Remark 9.2.3 and avoiding the combination of tile functions. If not found, the search continues on $\mathcal{G}'$. A found path is then bound to contain already configured tiles. Lines $13 - 21$ iterate over the path links and tiles until the first configured tile is found (lines $16 - 20$). Then, the outputs (link l' and reached tile h') of the already deployed function are calculated (line 17). The subroutine is then recursively called to find a path from h' to the intended receiver, while also avoiding links already visited.

Returning to the workflow of KpConfig (Algorithm 2), line 17 is reached with a maximum of K_e found paths for the studied pair. After accounting for any surplus paths that have remained from preceding pairs (line 17), an objective-driven selection of paths takes place at line 18 via the *FilterPathsByObj* subroutine (Algorithm 6).

FilterPathsByObj considers a power maximization *or* a QoS optimization objective, since these cannot be generally combined. The path selection for power transfer maximization is straightforward: the paths are sorted by total power and the top ones are selected. For a QoS optimization, the paths are first ordered by total delay. Then, a sliding window-based selection takes place, keeping the path subset that maximizes relation (9.31) subject to (9.32). Notice this approach does not take into account the signal interference (relation (9.34)). Instead, an eavesdropping objective can be added to all user pairs, thus naturally minimizing or mitigating interference as well.

KpConfig (Algorithm 2) proceeds to *Deploy* the selected paths at line 20, updating the f_h status in $\mathcal{G}$ as well. The deployment takes place by setting collimation functions, f_h^{COL}, at the first and last tile of the path, and steering functions, f_h^{STR}, to intermediate ones. Already configured tiles are not affected. Required modifiers, m_h^{PHA} or m_h^{POL} are applied afterwards sequentially, at the tile that yields the smallest effect on the total path gain $\prod g$.

At line 22, KpConfig (Algorithm 2) takes action to block unauthorized users, contained in the *blockedpairs* set. The process is given in the subroutine *DeployBlocks* (Algorithm 7). Since the objective is to block a single user from all others, we proceed to iterate over the user links of u_{TX} and fully absorb the emissions at each one. Notice that the immediately affected tiles may have a deployed function. Thus, in the general case, *DeployBlocks* searches for a path from u_{TX} to an arbitrary node (random user u_{RX} or any random tile). Once found, the path is iterated over and a full absorption function, f_h^{ABS}, is deployed at the first unconfigured tile h.

Finally, KpConfig (Algorithm 2) concludes at line 23 by setting all unused tiles in $\mathcal{G}$ to absorb from an arbitrary direction (e.g., from the tile surface normal). Thus, parasitic function outputs within PWE can be attenuated. It is noted that this step can be omitted, e.g., in cases where limiting the total number of configured tiles is a concern.

9.4 ARTIFICIAL INTELLIGENCE-BASED CONFIGURATION OF PWEs

Apart from the presented deterministic approach for configuring PWEs, heuristics can also be used. The potential advantage is operation in the complete solution space, without simplification assumptions (e.g., the one EM function per tile of KpPath). Thus, in this section we present an approach based on machine learning algorithms, in particular, neural networks, to adaptively configure PWEs for a set of users [102].

The key-idea is that, since HyperSurface tiles can regulate the distribution of power within a space, they can be represented by nodes in a neural network, while the power distribution can be mapped to neural network links and their weights. The weights are optimized via custom feed-forward/back-propagation processes, and are interpreted into HyperSurface tile functionalities. The machine learning approach is shown to be intuitive in its representation and economic in its use of HyperSurface tiles, compared to related approaches [103].

Consider a PWE environment comprising a wireless transmitter (Tx), a receiver (Rx) and a set of walls-scatterers coated with Hyper-Surface tiles. This environment is illustrated in Fig. 9.5.

We assume that EM waves transmitted from the Tx impinge upon the first wall. Thus, each HyperSurface tile over wall$_1$ has its own "input" impinging power. Each tile can be tuned to split and redirect its impinging power to other walls and tiles at their line-of-sight (LOS). These LOS links are shown as dashed arrows in Fig 9.5. Finally, the Rx receives part of the originally transmitted power via tiles in his LOS. The received power from each Rx link can be considered as the "output" of the propagation process. The "ideal output" represents lossless propagation, i.e., receiving the total of the transmitted power. The distribution over the Rx links respects the corresponding receiving gains of the Rx device, as derived by the Rx antenna pattern and the active MIMO configuration. The ideal and received outputs can be

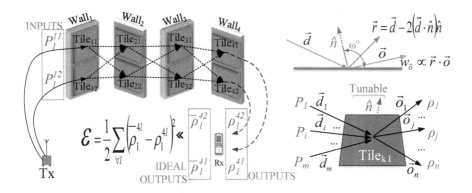

Figure 9.5 Conceptual representation of wireless propagation as a neural network.

compared, in order to obtain a metric of deviation, $\mathcal{E}$, as shown in Fig. 9.5.

We proceed to study the inputs/outputs of a single tile and their relation. Following the HyperSurface model described in the previous section, the HyperSurface tile steering can be modeled as virtually rotating the tile surface normal unit vector $\hat{n}$. The steering effect is shown in the top-right inset of Fig. 9.5. The reflected wave direction, $\vec{r}\left(\hat{n}, \vec{d}\right) = \vec{d} - 2\left(\vec{d} \cdot \hat{n}\right)\hat{n}$ [62], for a wave impinging from a direction $\vec{d}$ can be altered by tuning $\hat{n}$, which is achieved by setting the Hyper-Surface active elements accordingly. This information is provided by the HyperSurface manufacturer, e.g., in the form of a lookup table.

A reflection $\vec{r}\left(\hat{n}, \vec{d}\right)$ may not coincide exactly with a given inter-tile, "outgoing" link direction $\vec{o}$. Thus, we proceed to make a *heuristic* assumption that the power fraction, $w_{\vec{o}}$, over an "outgoing" link $\vec{o}$ is:

$$w_{\vec{o}}\left(\hat{n}, \vec{d}\right) = \frac{\max\left\{\vec{r}\left(\hat{n}, \vec{d}\right) \cdot \vec{o}, 0\right\}}{\sum_{\forall \vec{o}} w_{\vec{o}}}, \tag{9.38}$$

i.e., the projection of $\vec{r}$ over $\vec{o}$ normalized across all outgoing links, assuming that $\vec{o}$ and $\vec{d}$ are unit vectors. A negative inner product in relation (9.38) implies that vectors $\vec{r}$, $\vec{o}$ do not face in the same direction and, thus, $w_{\vec{o}}$ should intuitively be zero.

Based on these remarks, we proceed to model the wireless propagation as a neural network with walls as layers and tiles as nodes, and define feed-forward and back-propagate processes as follows:

9.4.1 Feed-Forward

Consider the general case of tile$_{k,l}$, indexed and denoted as shown in Fig. 9.5. The power at each of its outgoing links is:

$$\rho_j^{(k,l)} = \sum_{\forall \vec{d}_i} w_{\vec{\partial}} \left(\hat{n}, \vec{d}_i \right) \cdot P_i^{(k,l)}, \tag{9.39}$$

which serves as input power for the next wall-layer, since HyperSurfaces can employ collimation functionalities to focus their total outgoing power to the next tile [103]. At the final wall-layer, $k = \kappa$, the metric of deviation, $\mathcal{E}$, is calculated as:

$$\mathcal{E} = \frac{1}{2} \sum_{\forall l} \left(\overline{\rho}_1^{(\kappa,l)} - \rho_1^{(\kappa,l)} \right)^2 = \frac{1}{2} \sum_{\forall l} \delta_l^2, \tag{9.40}$$

where $\overline{\rho}$ denotes ideal output power ratio, and $\delta_l = \overline{\rho}_1^{(\kappa,l)} - \rho_1^{(\kappa,l)}$.

9.4.2 Back-Propagation

Once the walls-layers have been iterated over until the final layer, each tile-node at each wall-layer (in reverse order) deduces the effect of its currently active $\hat{n}$ vector to the deviation $\mathcal{E}$, and updates $\hat{n}$ to a new value $\hat{n}_*$. For simplicity we will focus on the 2D case where $\hat{n}$ is defined by a single angle ω, as shown in Fig. 9.5 (top-right). The corresponding update rule from ω to ω_* at tile (k, l) follows the generalized delta rule [59]:

$$\omega_*^{(k,l)} = \omega^{(k,l)} - \eta \cdot \left(\frac{\partial \mathcal{E}}{\partial \omega} \right)^{(k,l)} \cdot \mathcal{S}^{(k,l)}, \tag{9.41}$$

where $\eta \in (0, 1]$ is the network's learning rate, and $\mathcal{S}^{(k,l)}$ is a factor denoting the significance of tile (k, l) with regard to the propagation. For the final wall-layer $k = \kappa$ we define that $\mathcal{S}^{(\kappa,l)} = \delta_l$ (i.e., its deviation from the local ideal output). For $k \neq \kappa$ we define it as the total power impinging on the tile, i.e., $\mathcal{S}^{(\kappa,l)} = \sum_{\forall i} P_i^{(k,l)}$.

Finally, we provide the following formulas for $\left(\frac{\partial \mathcal{E}}{\partial \omega} \right)^{(k,l)}$ per indexed layer, omitting the proofs:

$$\left(\frac{\partial \mathcal{E}}{\partial \omega}\right) \overset{(\kappa,l)}{=} \delta_l \cdot \frac{\partial \rho_1^{(\kappa,l)}}{\partial \omega^{(\kappa,l)}}$$

$$\left(\frac{\partial \mathcal{E}}{\partial \omega}\right) \overset{(\kappa-1,l)}{=} \underbrace{\sum_{\forall j} \frac{\partial \rho_j^{(\kappa-1,l)}}{\partial \omega^{(\kappa-1,l)}}}_{e_j} \cdot \underbrace{\delta_j \cdot w_{\overrightarrow{o_j}}^{(\kappa-1,l)}}_{a_j} = \mathbf{e}^{(\kappa-1,l)} \cdot \mathbf{a}^{(\kappa-1,l)}$$

$$\dots \tag{9.42}$$

$$\left(\frac{\partial \mathcal{E}}{\partial \omega}\right) \overset{(k,l)}{=} \sum_{\forall j} \frac{\partial \rho_j^{(k,l)}}{\partial \omega^{(k,l)}} w_{\overrightarrow{o_j}}^{(k,l)} \left(\mathbf{1} \cdot \mathbf{a}^{(j,l+1)}\right),$$

where $\mathbf{e}$ and $\mathbf{a}$ are helping vectors defined as shown above, and $\mathbf{1}$ is an all-ones vector with size equal to $\mathbf{a}$. $\mathbf{e}$ is only used for facilitating the definition of $\mathbf{a}$ in relation (9.42) and has no further use, while $\mathbf{a}$ is updated recursively at each node. Its elements are the link weight products, for all paths leading from a right-layer neighbor to any output layer node.

The described feed-forward/back-propagate cycles can be executed in an online or offline manner, until the deviation $\mathcal{E}$ stabilizes, reaches an acceptable level or an allocated computational time window expires. At this point, the PWE controller simply deploys EM functionalities at each tile (k, l), matching the attained $\omega^{(k,l)}$ (and, thus $\hat{n}^{(k,l)}$) values. This trait makes the proposed approach a directly interpretable neural network configurator for PWEs. We denote this proposed approach as *NNConfig* and proceed to evaluate it via full PWE simulations.

9.5 THE NOVEL PWE POTENTIAL IN COMMUNICATION QUALITY, CYBERSECURITY, AND WIRELESS POWER TRANSFER

We proceed to evaluate the performance of *KpConfig* and *NNConfig* in a set of floorplans, users and objectives. We seek to demonstrate and visualize compliance to performance objectives, and deduce the performance gains in comparison with natural propagation (non-PWE).

The evaluation is based on a novel tool, implementing the process described in [103], developed specifically for simulating PWEs. The tool receives a floorplan, a set of configured tiles, as well as users' locations and user transmission characteristics. Subsequently, it essentially executes the process of Algorithm 1. Its simulation output is the power-delay profile for each user, which is then used for quantifying the compliance with the set objectives. The tool is implemented in JAVA over the AnyLogic platform [164]. The latter is chosen due to its strong visualization capabilities, and its versatility in combining simulation models (discrete events, agents, continuous processes)

Table 9.3 Persistent Simulation parameters.

Ceiling Height	$3\,m$
Tile Dimensions	$1 \times 1\,m$
Tile Functions	f_h^{COL}, f_h^{STR}, f_h^{ABS}
Non-HSF surfaces	Perfect reflectors
User scattering model	Blocking spheres, radius $0.5\,m$
Frequency	$2.4\,GHz$
Tx Power	$-30\,dBm$
Antenna type	Single a^o-lobe sinusoid, pointing at ϕ^o, θ^o (cf. Fig. 9.6)
Max ray bounces	50
Min considered ray power	-250 dBm

to describe a complex system. The developed tool allows for realistic simulation of PWE, supporting the following features:

■ Using any EM tile profile as input, supporting any physical implementation approach. The EM profiles can be input from Computational EM packages, or real measurements in a well-defined format.

■ Customized antenna radiation patterns with MIMO support.

■ Allowing for controllable reflection, refraction and diffraction of EM waves, along with any described metasurface functionality.

■ Flexible scenario creation via a 3D Graphical User Interface, allowing for varying tile topologies, dimensions, floorplan, user placement and roles, user mobility trajectories, and partially coated environments.

■ Flexible user objective definition mechanism, readily supporting all described objectives and allowing for creating custom ones.

■ Multi-cast and broadcast support.

■ Script-able and automate-able, parallelized simulations, allowing for automated parameter optimization, parameter sensitivity, Monte Carlo experiments, and general-purpose heuristic optimization runs [164].

Table 9.3 summarizes the persistent parameters across all subsequent tests. The following notes are made:

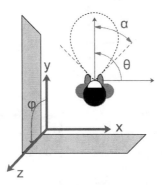

Figure 9.6 User location and beam orientation coordinate system. The origin is at the lower left corner of each floorplan.

- The user positions and radiation characteristics are considered to be known. These can be deduced by a user localization scheme specific for PWEs [170] or a third party scheme.

- All tiles have the same EM profile (cf. Remark 9.2.3), which is defined as follows: all impinging waves are attenuated by a constant factor of $g = 1\%$ of the carried power, for any type of deployed function. While simplified, this profile can represent existing EM models [196], when the meta-atom resolution is infinite [182].

- The considered tile functionalities include collimation, steering and absorption, while modifiers are not taken into account. As previously described, collimation is the effect of aligning EM waves to propagate over a flat front, rather than to dissipate over an ever-growing sphere. Thus, the path loss between two tiles in a PWE is not subject to the $\propto 1/d^2$ rule, d being their distance (cf. Fig. 9.3) [95]. This rule is only valid for the first impact, i.e., from the transmitter to its LOS tiles. The antenna aperture effect and gain are taken into account as usual.

- The antenna patterns of the transmitter and the receiver are simplified as single-lobe sinusoids, with the characteristics and θ, ϕ orientation shown in Fig. 9.6 and Table 9.3. In some scenarios, we assume that the mobile devices have beamforming capabilities and are able to turn the antenna lobe towards the ceiling, in conjunction with the mobile devices' gyroscopes.

Finally, all walls, ceilings and floors are considered as fully coated with HyperSurfaces, unless otherwise stated (defined per case).

Table 9.4 Scenario setup for the multi-user multi-objective showcase.

USER: POSITION, α,ϕ
0: $[1.0,10.0,1.0],10.0^{o},0^{o}$
1: $[8.5,11.0,1.0],80.0^{o},0^{o}$
2: $[15.6,11.5,1.0],10.0^{o},0^{o}$
3: $[11.0,5.4,1.0],10.0^{o},0^{o}$
4: $[6.0,6.0,1.0],10.0^{o},0^{o}$
5: $[8.5,1.6,1.0],60.0^{o},0^{o}$

PAIR: OBJECTIVE
$0{\to}2$: MAXPOWER,EAVESMIT[[ALL]]
$1{\to}3$: MAXPOWER,EAVESMIT[[ALL]]
$1{\to}4$: MAXSIR,EAVESMIT[[ALL]]
$5{\to}\times$: BLOCK

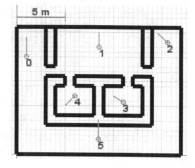

9.5.1   Multi-User Multi-Objective Showcase

This scenario showcases the PWE performance for three communicating pairs and one unauthorized user in a floorplan, as described in Table 9.4. The pairs have various different objectives, while there is also a multicast group (1 to 3, 4). Moreover, the antenna characteristics vary across the users.

The achieved wireless propagation, with and without PWE, is illustrated in Fig. 9.7. The PWE, configured via the proposed KPCONFIG, customizes the EM propagation to uphold all objectives (Fig. 9.7(a)). Specifically, the propagation for pair $0 \to 2$ follows a short route within the floorplan, to achieve maximal received power, while also avoiding all potential eavesdroppers. The pair $1 \to 3$ is treated similarly, i.e., maximizing the received powered without QoS concerns stemming from the length of each ray. For the pair $1 \to 4$, however, QoS is taken into account, resulting into a selection of rays that have similar total length, as shown in Fig. 9.7(a). Notice that, once the multicast group is served, the environment is tuned to absorb all unemployed emissions from user 1 (black lines). Finally, user 5 has its EM emissions absorbed, and is thus blocked from accessing all other users.

Natural propagation (i.e., non-PWE) within the same floorplan is shown in Fig. 9.7(b). While some pairs achieve a degree of connectivity (e.g., $1 \to 4$), the propagation is expectedly chaotic. The white lines denote stray rays within the floorplan, which eventually attenuate and

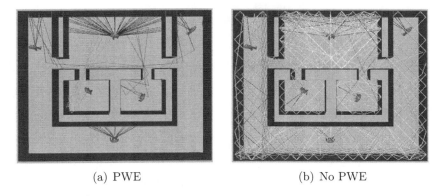

      (a) PWE   (b) No PWE

Figure 9.7   Propagation with and without PWE.

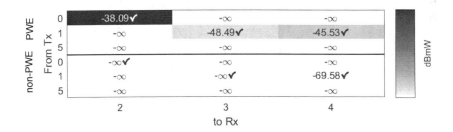

Figure 9.8   Comparative power reception with and without PWE. The 'Check' mark denotes intended pair connectivity.

disappear without reaching a user. Notice that the floorplan design, coupled with the highly directional antenna lobes of some users (e.g., $\alpha = 10^o$) naturally secludes users and hinders connectivity.

The received power levels are shown in Fig. 9.8. PWE achieves objective-compliant connectivity with a relatively small loss, ranging from ~ 8 to 15 dBmW. On the other hand, the non-PWE case achieves connectivity only for pair $1 \rightarrow 4$, with a considerably high loss of ~ 30 dBmW. It is noted that the PWE results of Fig. 9.8 are calculated as the total-power-minus-interference. The classification of a ray as interfering factors for the MIMO labeling of the user links is produced by KPCONFIG. (The interference was zero in all cases). In the non-PWE case, the interference classification does not consider any MIMO labeling, since there is not an automatic way of setting it.

Table 9.5 Scenario setup for the Doppler effect mitigation showcase.

USER: POSITION, α,ϕ
0: $[17.0,3.2,1.0],30.0^{o},90.0^{o}$
1: $[2.0,1.7,1.0],30.0^{o},90.0^{o}$

PAIR: OBJECTIVE
$1 \rightarrow 0 : \{$MAXPOWER,EAVMIT$[[0]],$ DOPPLERMIT$[trajectory]$ ″

9.5.2 Doppler Effect Mitigation Showcase

We proceed to examine the mitigation of Doppler Effects, using the setup described in Table 9.5. Both users have their antennas pointing upwards (i.e., $\phi = 90^{o}$). User 0 moves along a trajectory, and KPCONFIG proceeds to filter its user links, selecting and employing those that are most perpendicular to the trajectory. One or more paths can be established to connect the pair, using one or more of the user 0 links. We are interested in keeping the maximal deviation from perpendicularity bounded; the maximal angle formed by each user link and the trajectory must not exceed $90^{o} \pm 10^{o}$. If no user link fulfills this criterion, the link with the minimal deviation is selected.

The user links selected across the trajectory are illustrated in Fig. 9.9 as white lines. As the user moves, he is served by the tile nearest to him. This tile remains active for a few steps and is then succeeded by another one. When a user is exactly below each active tile, its deviation from perpendicularity becomes minimal. The deviation then increases until the user is positioned exactly below the next active tile. This effect is shown in Fig. 9.10(a), which also showcases the effect of tile placement across the trajectory. For the first 50 steps, the user follows a skewed direction with regard to the regular grid of the ceiling tiles (cf. the inset of Table 9.5). As such, the tile selection and deviation does not exhibit a well-defined pattern. For steps $50 - 175$, the trajectory is aligned with tile centers at the ceiling, exhibiting the described periodicity in the perpendicularity deviation. The pattern changes again for the last, skewed part of the trajectory.

Figure 9.10(b) shows the received power across the trajectory. It is worth noting that the received power drops when the deviation from perpendicularity spikes. Due to the user link selection process described above, only one connecting path may be active in these cases. This naturally limits the received power as well. Intermediate points in the

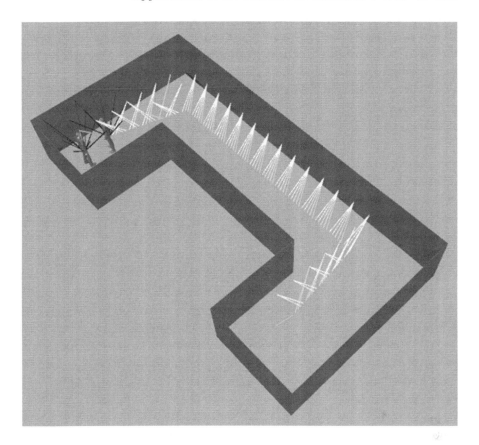

Figure 9.9 Selected user links for Doppler effect mitigation across a trajectory.

trajectory allow for 2 or 3 concurrent paths, increasing the total carried power.

Finally, notice that the upwards directivity of the antennas means that there is no LOS component across the trajectory. Moreover, the use of collimation tile functions means that power losses are mainly encountered over the user links (cf. Fig. 9.3). However, the user link effects are static in this scenario, due to the upwards directivity of the antennas. Thus, the received power is mainly affected by the paths deployed between users 0 and 1.

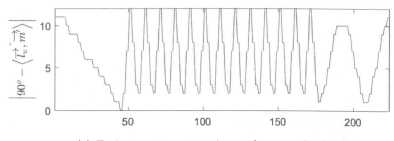

(a) Trajectory-to-propagation path perpendicularity.

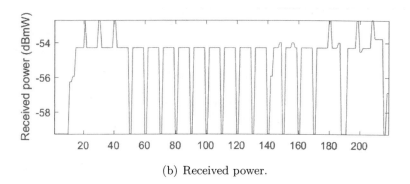

(b) Received power.

Figure 9.10 Trajectory-to-propagation path perpendicularity and receiver power across the user trajectory. (The x-axis represents steps of 12.5 cm across the trajectory).

9.5.3 User Capacity and Stress Test

The final test studies the capacity of a PWE, subject to a configuration algorithm such as KPCONFIG. In other words, we seek to understand the conditions under which the PWE performance gains diminish to non-PWE levels.

The scenario setup is given in Table 9.6. It comprises eight communicating pairs and three system stress levels. In the first level all transmitters have antennas with $\alpha = 50^o$, while the HyperSurface coating is full over walls, floor and ceiling (494 tiles total). The second stress level is the same but with $\alpha = 80^o$, naturally leading to more emitted rays to handle per transmitter. The last stress level is similar to the second, but only the ceiling is HyperSurface-coated (169 tiles total). Thus, the ray control points are fewer. Notice that partial coatings can be accounted for in KPCONFIG by passing a graph input $\mathcal{G}$ that already

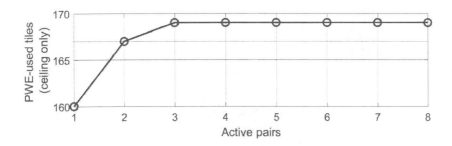

Figure 9.11 Tiles activated (out of 169 total) as the number of user pairs increases, corresponding to the case of Fig. 9.13(c).

contains tiles with $f_h \neq \emptyset$. In other words, areas without tile coating are marked as having virtual tiles, pre-configured with functions describing natural propagation. Finally, all user antennas are pointing upwards, yielding only NLOS components for all pairs.

Figure 9.12 presents the received power and interference among users per stress level, as well as the number of rays impinging per tile in boxplot form. (All tiles are considered, virtual or real, active, or inactive). In the first stress level, the PWE is configured by the KPCONFIG almost ideally (cf. Fig. 9.12(a)). The intended pairs are connected and well-separated. Minor interference is noticed explained as follows: with $\alpha = 50°$, the affected tiles per transmitter marginally overlap. KPCONFIG configures roughly 20% of the available tiles ($101/494$). The tile re-use remains at very low levels, with 1 ray per tile being the most common case, as intended. The non-PWE does not exhibit a well-defined pattern and yields approximately 30 dBmW of extra loss per pair, and complete disconnection for three pairs.

The second stress level follows the same pattern but with decreased performance in both PWE and non-PWE (Fig. 9.12(b)). The PWE case continues to show good pair connectivity (notice the diagonal of intended pairs), but interference has increased as expected. KPCONFIG configures almost 50% of the available tiles ($256/494$). The same effect is evident in the non-PWE case, validating the presence of increased system stress. The tile re-use is kept low in the PWE case.

The final stress case shows a collapse in the PWE performance, making it almost indistinguishable from the non-PWE case (Fig. 9.13(c)). KPCONFIG uses all available tiles ($169/169$). The high stress level is also evident from the tile reuse, which is almost identical for both PWE and non-PWE. In Figure 9.11 we execute the same

Table 9.6 Scenario setup for the PWE stress test.

USER: POSITION, α,ϕ
0: $[2.5,10.0,1.0],\{50^o,80^o\},90.0^o$
1: $[5.0,10.0,1.0],\{50^o,80^o\},90.0^o$
2: $[7.5,10.0,1.0],\{50^o,80^o\},90.0^o$
3: $[10.0,10.0,1.0],\{50^o,80^o\},90.0^o$
4: $[2.5,7.5,1.0],\{50^o,80^o\},90.0^o$
5: $[5.0,7.5,1.0],\{50^o,80^o\},90.0^o$
6: $[7.5,7.5,1.0],\{50^o,80^o\},90.0^o$
7: $[10.0,7.5,1.0],\{50^o,80^o\},90.0^o$
8: $[2.5,5.0,1.0],180.0^o,90.0^o$
9: $[5.0,5.0,1.0],180.0^o,90.0^o$
10: $[7.5,5.0,1.0],180.0^o,90.0^o$
11: $[10.0,5.0,1.0],180.0^o,90.0^o$
12: $[2.5,2.5,1.0],180.0^o,90.0^o$
13: $[5.0,2.5,1.0],180.0^o,90.0^o$
14: $[7.5,2.5,1.0],180.0^o,90.0^o$
15: $[10.0,2.5,1.0],180.0^o,90.0^o$
PAIR: OBJECTIVE
$(i) \to (15-i)$, $i = 0\ldots7$:
MAXPOWER,EAVMIT[[ALL]]

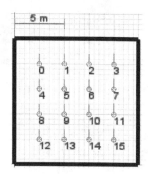

measurements as in Fig. 9.12, but by activating the transmitters in a serial fashion. Evidently, after a single transmitter is activated, almost all of the PWE tiles are used. This is natural, given that an antenna lobe of $\alpha = 80^o$ pointing upwards affects almost all of the ceiling tiles. Subsequent transmitters find no available free tiles to handle them, leading to uncontrolled propagation. Thus, at this point the user capacity of the PWE, *subject to* KPCONFIG *as the configuration scheme*, has been exceeded.

9.5.4 Evaluation of Neural Network-Based PWE Heuristics

We consider a Tx-Rx pair in the floorplan and setup shown in Table 9.7. It comprises three walls that the Tx emissions must sequentially and necessarily hit to reach the Rx. We seek to tune each SDM tile in this setup so as to maximize the received power at the Rx. Note that the

Table 9.7 Floorplan, user location and beam orientation coordinate system. The origin is at the lower left corner of the floorplan.

USER: POSITION, α, ϕ, θ
0: [2.5, 7.5, 0.5], 40°, 0°, 0°
1: [7.5, 7.5, 0.5], 40°, 0°, 180°
COMM. PAIR: OBJECTIVE
$0 \rightarrow 1$: MAX. RECEIVED POWER

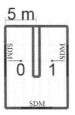
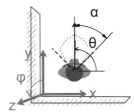

setup corresponds to 2D ray propagation in a 3D space (the tile size and floor height are equal).

The *NNConfig* neural network representation of the simulation setup is shown in Fig. 9.13. The input to each tile of the first layer is the normalized portion of power impinging upon it via the links of user 0. Since the objective is to transfer all emitted power to user 1, these inputs can be *virtual* (i.e., not equal to the actual impinging power distribution). Thus, each input is set to 20 % of the total emitted power, for each of the five tiles in the first layer. Using the same principle, the ideal output is set to 20 % of the total power to reach the receiver. Since the inputs and outputs remain the same at every feed-forward/back-propagate cycle, a high learning rate is selected ($\eta = 0.95$). The initial ω angle values per tile are randomized in the range $[-90^o, 90^o]$. This leads to the formation of the untrained network of Fig. 9.14(a). Finally, the termination criterion is to reach a root mean square error (RMSE) of less that 10^{-3}, which is achieved at approximately 2000 feed-forward/back-propagate cycles.

The trained network is shown in Fig. 9.14(b). Notably, *NNConfig* finds a solution where only one tile is activated in the middle layer. This is the natural outcome of reinforcement learning around an "impactful" tile, i.e., one already managing a considerable part of the total impinging power. Thus, *NNConfig* is economic in terms of used tiles, which can be beneficial in terms of power expenditure and supported user capacity.

Visualizations of the ray-traced outcome per each compared scheme are given in Fig. 9.14. Regular propagation supports only symmetric (specular) ray reflection. As a result, it cannot steer $\sim$ 60 % of the emitted power to the receiver; since three out of five rays (to the front and right of the transmitter) are directly reflected to absorbing areas.

Table 9.8 Received Power per Approach.

Regular propagation	$-70.71\ dBmW$
KPCONFIG	$-49.87\ dBmW$
NNCONFIG	$-49.83\ dBmW$

KpConfig works as intended and finds a solution that routes all power to the receiver. However, it uses all tiles in the middle wall, reaching full PWE capacity. *NNConfig* uses just one tile in the middle, in perfect match with the neural network abstraction in Fig. 9.14(b).

The received power per scheme is given in Table 9.8. All three propagation approaches are subject to receiving antenna aperture losses [103]. Regular propagation performs worse for two reasons. First, it loses ∼ 60 % of the emitted power for the reasons described above. Second, it does not support collimation and focusing capabilities and is, thus, subject to the $1/r^2$ power attenuation rule [103]. On the other hand, both *KpConfig* and *NNConfig* behave similarly in terms of received power, since they successfully groom and route all power from the transmitter to the receiver.

Discussion and future work

NNConfig exhibits potential for configuring PWEs in a tile-economic manner. It can minimize the number of tiles required for serving communication objectives. Moreover, it follows a representation of the PWE setting that is simple, intuitive and directly interpretable. Future work is directed towards extending the neural network representation to multiple users and multiple objectives. To this end, a promising direction is its combination with *KpConfig* in a hierarchical approach, where *KpConfig* coarsely routes EM waves at the level of wall orderings, and *NNConfig* deduces the exact EM functionality per tile.

Novel directions study the interplay between network latency and its effect on HyperSurface functionality efficiency [96]. When a user device moves within a space, the orientation of its reception and transmission antennas necessarily changes. As such, the programmable wireless environment may be misaligned to actual position of the user devices. The study of showcased this effect for the case of beam focusing and steering, in a partially deployed programmable wireless environment. It was shown that the PWE efficiency fades quite fast around the intended/perceived user position. Moreover, a novel misalignment mitigation framework was proposed based on the concept of fuzzy beam forming. According to this technique, the HyperSurface-departing wavefront is shaped to cover not only the perceived user location, but rather an area around it, defined by a lax prediction based on the user mobility pattern. The fuzzy beamforming is achieved by assigning HyperSurface elements to discretized, served positions in space. Classic scheduling with weights approaches have been shown to be effective in yielding the assignment in an optimal manner [97, 98, 105].

Future studies will generalize this direction, and study the interplay of the network components with the PWE efficiency in a general frame.

9.6 CONCLUSION

Programmable wireless environments enable the customization of wireless propagation within them. The present work presented proposed schemes to configure such environments for serving multiple users and multiple objectives. Exemplary objectives include security against eavesdropping, mitigating Doppler effects, power transfer and signal-to-interference maximization. One scheme, the KPCONFIG, employed a novel, graph-based model of programmable environments, which transforms performance objectives to path search problems, while taking into account core Physical restrictions. KPCONFIG was extensively evaluated in a novel tool for simulating programmable wireless environments, yielding significant performance gains over regular propagation and reaching insights on the user capacity of PWEs. Moreover, an interpretable neural network-based approach for configuring the behavior of tiles in such environments, the NNConfig, was also presented. Tiles and inter-tile power flow were mapped to neural network nodes and links respectively, and custom feed-forward and back-propagate processes were introduced. Evaluation via the same ray-tracing tool showed performance potential at the level of state-of-the-art solutions, with a distinct gain in reducing the total number of tiles required to be activated.

ACKNOWLEDGMENTS

This work was partially funded by the European Union's Horizon 2020 research and innovation program, Future and Emerging Topics (FET Open) under grant agreement No 736876.

Algorithm 3 The "Most Distant Pair First" resource sharing policy for PWEs.

1: **procedure** $blockedpairs$, $sortedpairs$, N_e, K_e:

2:   $MDFPolicy(\ \mathcal{G}, \{\langle u_{\mathrm{TX}}, u_{\mathrm{RX}}, \mathrm{OBJ} \rangle\})$

3:     $n_u \leftarrow \{0, \ldots, 0\};$ //for all users.

4:     $sortedpairs \leftarrow \{\emptyset\};$

5:     $blockedpairs \leftarrow \{\emptyset\};$

6:     $v_e \leftarrow \{\emptyset\};$

7:     $K_e \leftarrow \{\emptyset\};$

8:     $N_e \leftarrow \{\emptyset\};$

9:     **for** $e : \langle u_{\mathrm{TX}}, u_{\mathrm{RX}}, \mathrm{OBJ} \rangle$ **in** $\{\langle u_{\mathrm{TX}}, u_{\mathrm{RX}}, \mathrm{OBJ} \rangle\}$

10:       **if** $\mathrm{OBJ} = \mathrm{BLOCK}$// u_{RX} *is arbitrary.*

11:         $blockedpairs \leftarrow blockedpairs + e;$

12:         **continue;**

13:       **else;**

14:         $sortedpairs \leftarrow sortedpairs + e;$

15:       **end if**

16:       $K_e \leftarrow \min \{\|\mathcal{L}_{u:u_{\mathrm{TX}}}\|, \|\mathcal{L}_{u:u_{\mathrm{RX}}}\|\};$

17:       $\{\overrightarrow{p}\}_e \leftarrow \mathrm{KSHORTESTPATHS}\,(\mathcal{G}, K_e, u_{\mathrm{TX}}, u_{\mathrm{RX}});$

18:       $v_e \leftarrow \mathrm{MEAN}\,(\{\mathrm{DELAY}\,(\overrightarrow{p})\}_e);$

19:       $n_{u_{\mathrm{TX}}} \leftarrow n_{u_{\mathrm{TX}}} + \|\{\overrightarrow{p}\}_e\|;$

20:       $n_{u_{\mathrm{RX}}} \leftarrow n_{u_{\mathrm{RX}}} + \|\{\overrightarrow{p}\}_e\|;$

21:     **end for**

22:     $sortedpairs \leftarrow \mathrm{SORTDESC}\,(sortedpairs, v_e);$

23:     **for** $e : \langle u_{\mathrm{TX}}, u_{\mathrm{RX}}, \mathrm{OBJ} \rangle$ **in** $sortedpairs$

24:       $N_e \leftarrow \max \left\{ \min \left\{ \left\lfloor \frac{\|\mathcal{L}_{u:u_{\mathrm{TX}}}\|}{n_{u_{\mathrm{TX}}}} \right\rfloor, \left\lfloor \frac{\|\mathcal{L}_{u:u_{\mathrm{RX}}}\|}{n_{u_{\mathrm{RX}}}} \right\rfloor \right\}, 1 \right\};$

25:     **end for**

26: **end procedure**

Algorithm 4 Obtaining links in conflict with eavesdropping and Doppler effect mitigation objectives.

1: **procedure** $\mathcal{L}_a$: FILTERLINKSBYOBJ($\mathcal{G}$, OBJ)

2: $\quad \mathcal{L}_a \leftarrow \{\}$;

3: $\quad$ **switch** OBJ:

4: $\quad\quad$ **case** MITIGATEEAVESDROP: *//Inputs of (9.36) implied.*

5: $\quad\quad\quad \mathcal{L}_a \leftarrow \mathcal{L}_a +$ relation (9.36);

6: $\quad\quad$ **case** MITIGATEDOPPLER: *//Inputs of (9.37) implied.*

7: $\quad\quad\quad \mathcal{L}_a \leftarrow \mathcal{L}_a +$ relation (9.37);

8: $\quad$ **end switch**

9: **end procedure**

Algorithm 5 The process for finding a single path in a PWE.

1: **procedure** $\vec{p}$: *FindComplexPath*($\mathcal{G}, u_{\text{TX}}, u_{\text{RX}}, \mathcal{L}_a, paths$)

2:     $\mathcal{G}' \leftarrow \mathcal{G} \langle \{\mathcal{H}, u_{\text{TX}}, u_{\text{RX}}\}, \{\mathcal{L}_h, \mathcal{L}_{u_{\text{TX}}}, \mathcal{L}_{u_{\text{RX}}}\} - \mathcal{L}_a - paths \rangle$;

3:     $\mathcal{G}^* \leftarrow \mathcal{G}' - \langle \{\mathcal{H} | f_h = \emptyset\} \rangle$;

4:     $\vec{p} \leftarrow$ SHORTESTPATH $(\mathcal{G}^*, u_{\text{TX}}, u_{\text{RX}})$;

5:     **if** $\vec{p} \neq \emptyset$

6:         **return** $\vec{p}$; //*Path without deployed functions found.*

7:     **end if**

8:     $\vec{p}^* \leftarrow$ SHORTESTPATH $(\mathcal{G}', u_{\text{TX}}, u_{\text{RX}})$;

9:     **if** $\vec{p}^* = \emptyset$

10:        **return** $\emptyset$; //*No path exists*

11:     **end if**

12:     $\vec{p} \leftarrow \{\}; h' \leftarrow \emptyset$;

13:     **for** $\langle l, h \rangle$ **in** $\vec{p}^*$

14:       **if** $f_h = \emptyset$

15:         $\vec{p} \leftarrow \vec{p} + l$;

16:       **else**

17:         $\langle l', h' \rangle \leftarrow f_h (l)$;

18:         $\vec{p} \leftarrow \vec{p} + l'$;

19:         **break**;

20:       **end if**

21:     **end for**

22:     $\vec{p}^* \leftarrow$ SELF$(\mathcal{G}, h', u_{\text{RX}}, \mathcal{L}_a, paths - \vec{p})$;

23:     **if** $\vec{p}^* = \emptyset$

24:        **return** $\emptyset$;

25:     **else**

26:        **return** $\vec{p} + \vec{p}^*$;

27:     **end if**

28: **end procedure**

Algorithm 6 Objective-oriented propagation path selection.

1: **procedure** *selpaths:*

2:     *FilterPathsByObj*(*paths,maxpaths,*obj)

3:         **switch** OBJ

4:           **case** MAXPOWER: *//Inputs of (9.30) implied.*

5:             *paths* ← SORTDESC (*paths*); *// by* $\mathbf{P} \cdot \prod g_h$.

6:             *paths* ← KEEPTOPN (*paths, maxpaths*);

7:             **break**;

8:           **case** MAXSIR: *//Inputs of (9.31) implied.*

9:             *paths* ← SORTASC (*paths*); *// by path* DELAY.

10:             **for** i=1:1:$\|paths\|$

11:               $\mathbf{P}_i^{TOT}$ ←   *//by rel. (9.31),(9.32).*

12:                 $\mathbf{P}^{TOT}\left(paths_{i,\min\{i+max\mathbf{paths},\|paths\|\}}\right)$

13:             **end for**

14:             i^{OPT} ← $arg\max\left\{\mathbf{P}_i^{TOT}\right\}$;

15:             $paths_{i^{OPT},\min\{i^{OPT}+max\mathbf{paths},\|paths\|\}}$;

16:             **break**;

17:         **end switch**

18:         *selpaths* ← *paths*;

19: **end procedure**

Algorithm 7 The process for blocking user access in PWE.

1: **procedure** DEPLOYBLOCKS($\mathcal{G},blockedpairs$)

2:     **for** $e : \langle u_{\text{TX}}, u_{\text{RX}}, \text{OBJ}\rangle$ in *blockedpairs*

3:         **for** $l_{u,h^*} \in \mathcal{L}_u | u = u_{\text{TX}}$

4:           $\overrightarrow{p}$ ← FINDCOMPLEXPATH $(\mathcal{G}, h^*, u_{\text{RX}})$;

5:           **for** $h \in \overrightarrow{p}$

6:             **if** $f_h = \emptyset$

7:               f_h ← f_h^{ABS};

8:               **break**;

9:             **end if**

10:           **end for**

11:         **end for**

12:     **end for**

13: **end procedure**

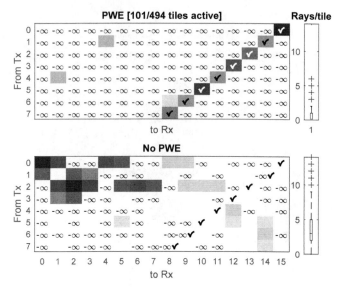

(a) Full tile coverage, $\alpha = 50^o$ for each transmitter.

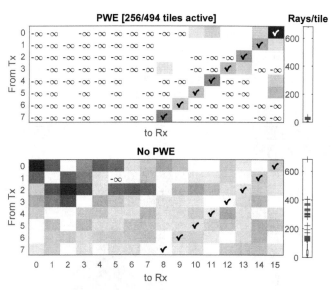

(b) Full tile coverage, $\alpha = 80^o$ for each transmitter.

Figure 9.12 Power reception and tile reuse for various tests of progressively increasing stress. The 'check' mark denotes intended pair connectivity. The range of values is common across all subplots: $[-106.94, -59.23]$ dBmW.

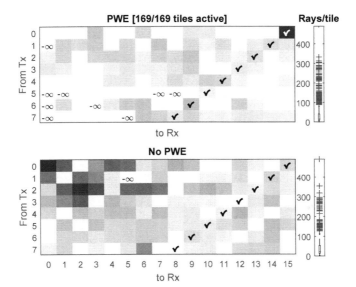

(c) Partial tile coverage (ceiling only), $\alpha = 80^{o}$ for each transmitter.

Figure 9.12 (*CONTINUED*)

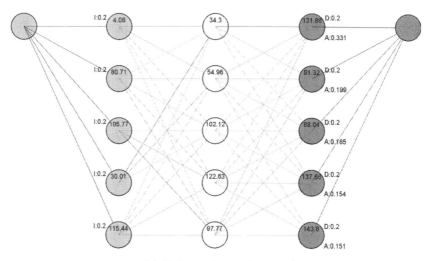

(a) Untrained neural network.

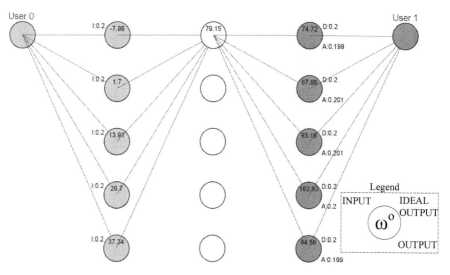

(b) Trained neural network after 2000 feed-forward/back-propagate cycles (learning rate $\eta = 0.95$, attained RMSE=$8 \cdot 10^{-4}$).

Figure 9.13 The neural network corresponding to the studied setup and its untrained/trained states.

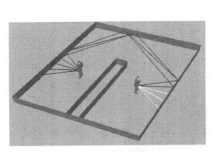

(a) Natural propagation.

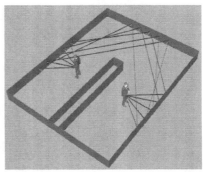

(b) Propagation via KPCONFIG.

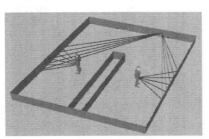

(c) Propagation via NNCONFIG.

Figure 9.14 Wireless propagation visualized for each approach.

CHAPTER 10

Epilogue

Christos Liaskos

Foundation for Research and Technology Hellas, 71110, Heraklion, Crete, Greece. Email: cliaskos@ics.forth.gr - corresponding author

This book provided the foundational knowledge and modeling of the Internet-of-Materials concept, covering the involved disciplines to an entry level. Starting from a high-level view of all the separate aspects comprising this novel concept, we delved into the foundational concepts of electromagnetism that permeate and describe the operation of metasurfaces and metamaterials, the metasurface foundations from the computer science aspect, introducing novel software classes and services, the hardware aspects of a HyperSurface and the network protocols that interconnect them to the Internet-of-Things ecosystem, the scaling laws of HyperSurfaces and, finally, concluded with showcasing the Internet of Metamaterials applied to the wireless communications case. In this chapter we conclude the book by discussing the further technological advancement goals for the HyperSurface concept.

Further research in programmable wireless environments can target the tile architecture and the inter-tile networking, the tile control software, and many applications such as mm-waves, D2D and 5G systems.

Regarding the tile architecture, the optimization of the dynamic meta-atom design constitutes a notable goal towards maximizing the supported function range of a tile. Ultra-wideband meta-atom designs able to interact concurrently with a wide variety of frequencies, e.g., from 1 to 60 GHz, constitute a notable research goal [151]. Formal tile sounding procedures need to be defined, i.e., simulation-based and experimental processes for measuring the supported functions and parameters per tile design. Additionally, the tile reconfiguration speed needs to be studied, in order to yield the adaptivity bounds of the programmable environments. In this sense, inter-tile networking protocols need to be designed to offer fast, energy-efficient wireless environment reconfiguration, supporting a wide range of user mobility patterns. Adaptation to user mobility can also target the mitigation of Doppler shift effects.

The HyperSurface control software needs to be optimized regarding its complexity, modularity, and interfacing capabilities. Low-complexity, fast configuration optimizers can increase the environment's maximum adaptation speed. Towards this end, both analysis-based and heuristic optimization processes need to be studied. Additionally, following the Network Function Virtualization paradigm [19], the various described and evaluated optimization objectives can be expressed in a modular form, allowing their reuse and combination. For example, the tiles may be configured to maximize the minimum received power within a room, subject to delay spread restrictions. Well-defined tile software interfaces can allow for a close collaboration with user devices and external systems. For instance, the power delay profile towards a user can be matched to the Multiple-Input Multiple Output arrangement of his device. It is noted that such joint optimization can be aligned to the envisioned 5G objectives of ultra-low latency, high bandwidth, and support for massive numbers of devices within an environment [19].

We note that the HyperSurface concept is applicable to any frequency spectrum and wireless architecture. Therefore, solving the corresponding path loss, fading, interference and non-line-of-sight problems in general using HyperSurfaces constitute promising research paths. Such directions can further focus on indoors and outdoors communication environments. In indoors settings, the HyperSurface tiles can cover large parts of the wireless environment, such as walls, ceilings, furniture and other objects and offer more precise control over electromagnetic waves. In outdoors settings, the HyperSurface tiles can be placed on key-points, such as building facades, highway polls, advertising panels, which can be utilized to boost the communication efficiency. In both settings, i.e., in indoors and outdoors, the automatic tile location and orientation discovery can promote the ease of deployment towards "plug-and-play" levels. Moreover, the joint optimization of antenna beam-forming and tile configurations needs to be studied, to achieve the maximum performance.

Studying the use of HyperSurfaces in mm-wave systems, 5G systems and THz communications is of particular interest. For example, mm-wave and THz systems are severely limited in terms of very short distances and LOS scenarios. The HyperSurfaces can mitigate the acute path loss by enforcing the lens effect and any custom reflection angle per tile, avoiding the ambient dispersal of energy and non-line-of-sight effects, extending the effective communication range. Dynamic meta-atoms that can interact with THz modulated waves need to be designed. This has been shown to be possible for graphene-based metasurfaces [87]. The tile sensing accuracy and re-configuration speed must also match the extremely high spatial sensitivity of THz communications, calling for novel, highly distributed tile control processes. Optical tile inter-networking is another approach to ensure that the tile adaptation service is fast enough for the THz communication needs.

Finally, the control of EM waves via HyperSurfaces can find applications beyond classic communications. EM interference constitutes a common problem in highly sensitive hardware, such as medical imaging and radar technology. In these cases, the internals of, e.g., a medical device can be treated as an EM environment, with the objective of canceling the interference to the imaging component caused by unwanted internal EM scattering. Such interference can be mitigated only up to a degree during the design of the equipment. Common discrepancies that occur during manufacturing can give rise to unpredictable interference, resulting into reduced equipment performance. However, assuming HyperSurface-coated device internals, interference can be mitigated or even negated, after the device manufacturing, via simple software commands.

REFERENCES

[1] BeagleBoard. https://beagleboard.org/.

[2] Clustering MQTT for high availability and scalability. https://vernemq.com/.

[3] CST Microwave Studio. https://www.3ds.com/products-services/simulia/products/cst-studio-suite/.

[4] Eclipse Mosquitto, an open source MQTT broker. https://mosquitto.org/.

[5] Espressif. https://www.espressif.com/.

[6] I2c - What is that? https://www.i2c-bus.org/.

[7] The leader in open source MQTT broker. https://www.emqx.io/.

[8] OpenWrt wireless freedom. https://openwrt.org/.

[9] Pin diodes. https://www.microwaves101.com/encyclopedias/pin-diodes.

[10] RabbitMQ. https://www.rabbitmq.com/.

[11] Resistor ladder. https://en.wikipedia.org/wiki/Resistorladder.

[12] TinyOS community. http://www.tinyos.net/.

[13] TinyOS: Operating system design for wireless sensor networks. https://www.fierceelectronics.com/iot-wireless/tinyos-operating-system-design-for-wireless-sensor-networks.

[14] Special issue on active and adaptive antennas. *IEEE Transactions on Antennas and Propagation*, 12(2), 1964.

[15] S. Abadal, T.-J. Cui, T. Low, and J. Georgiou. Programmable metamaterials for software-defined electromagnetic control: Circuits, systems, and architectures. *IEEE Journal on Emerging and Selected Topics in Circuits and Systems*, 10(1):6–19, 2020.

[16] S. Abadal, C. Liaskos, A. Tsioliaridou, S. Ioannidis, A. Pitsillides, J. Sole-Pareta, E. Alarcon, and A. Cabellos-Aparicio. Computing and communications for the software-defined metamaterial paradigm: A context analysis. *IEEE Access*, 5:6225–6235, 2017.

[17] O. Abari, D. Bharadia, A. Duffield, and D. Katabi. Enabling high-quality untethered virtual reality. In *14th USENIX Symposium on Networked Systems Design and Implementation, NSDI 2017*, pages 531–544, 2017.

[18] I. F. Akyildiz, C. Han, and S. Nie. Combating the distance problem in the millimeter wave and terahertz frequency bands. *IEEE Communications Magazine*, 56(June):102–108, 2018.

[19] I. F. Akyildiz, S. Nie, S.-C. Lin, and M. Chandrasekaran. 5G roadmap: 10 key enabling technologies. *Computer Networks*, 106:17–48, 2016.

[20] V. S. Asadchy, A. D\'\iaz-Rubio, S. N. Tcvetkova, D.-H. Kwon, A. Elsakka, M. Albooyeh, and S. A. Tretyakov. Flat engineered multichannel reflectors. *Phys. Rev. X*, 7(3):031046, 2017.

[21] N. Ashraf, T. Saeed, R. I. Ansari, M. Lestas, C. Liaskos, and A. Pitsillides. Feedback based beam steering for intelligent metasurfaces. In *2019 2nd IEEE Middle East and North Africa COMMunications Conference, MENACOMM 2019*, pages 1–6, 2019.

[22] C. A. Balanis. *Antenna Theory: Analysis and Design*. 3rd edition, Wiley Publishing, 2005.

[23] D. Bertsekas. *Nonlinear Programming*. Athena Scientific Belmont, 1999.

[24] K. Bi, W. Zhu, M. Lei, and J. Zhou. Magnetically tunable wideband microwave filter using ferrite-based metamaterials. *Appl. Phys. Lett.*, 106(17):173507, 2015.

[25] D. Bill and M. Watertown. Pin diode fundamentals. *Microsemi, Micronotes*. MicroNote Series 701.

[26] S. R. Biswas, C. E. Gutierrez, A. Nemilentsau, I.-H. Lee, S.-H. Oh, P. Avouris, and T. Low. Tunable graphene metasurface reflectarray for cloaking, illusion, and focusing. *Physical Review Applied*, 9(3):034021, 2018.

[27] F. Bourquin, G. Caruso, M. Peigney, and D. Siegert. Magnetically tuned mass dampers for optimal vibration damping of large structures. *Smart Mater. Struct.*, 23(8):085009, 2014.

[28] A. W. Brander and M. C. Sinclair. A comparative study of k-shortest path algorithms. In *Performance Engineering of Computer and Telecommunications Systems*, pages 370–379. Springer, 1996.

[29] P. E. Bulychev, A. David, K. G. Larsen, M. Mikucionis, D. B. Poulsen, A. Legay, and Z. Wang. UPPAAL-SMC: Statistical model checking for priced timed automata. In *Proceedings 10th Workshop on Quantitative Aspects of Programming Languages and Systems, QAPL 2012*, volume 85 of *EPTCS*, pages 1–16, 2012.

[30] S. N. Burokur, J. P. Daniel, P. Ratajczak, and A. D. Lustrac. Tunable bilayered metasurface for frequency reconfigurable directive emissions. *Appl. Phys. Lett.*, 97(6):064101, 2010.

[31] Z. Cao, X. Xiang, C. Yang, et al. Analysis of tunable characteristics of liquid-crystal-based hyperbolic metamaterials. *Liq. Cryst.*, 43(12):1753–1759, 2016.

[32] H. Chalabi, Y. Ra'di, D. L. Sounas, and A. Alu. Efficient anomalous reflection through near-field interactions in metasurfaces. *Physical Review B*, 96(7):075432, 2017.

[33] I. Chatzakis, L. Luo, J. Wang, et al. Reversible modulation and ultrafast dynamics of terahertz resonances in strongly photoexcited metamaterials. *Phys. Rev. B*, 86(12):125110, 2012.

[34] S. D. Chawade, M. A. Gaikwad, and R. M. Patrikar. Review of xy routing algorithm for network-on-chip architecture. *International Journal of Computer Applications*, 43(21):48–52, 2012.

[35] H.-T. Chen, J. F. O'hara, A. K. Azad, et al. Experimental demonstration of frequency-agile terahertz metamaterials. *Nat. Photonics*, 2(5):295–298, 2008.

[36] H. T. Chen, A. J. Taylor, and N. Yu. A review of metasurfaces: Physics and applications. *Reports on Progress in Physics*, 79(7), 2016.

[37] K. Chen, Y. Feng, F. Monticone, et al. A reconfigurable active Huygens' metalens. *Adv. Mater.*, 29(17):1606422, 2017.

[38] D. Chicherin, S. Dudorov, D. Lioubtchenko, et al. MEMS-based high-impedance surfaces for millimeter and submillimeter wave applications. *Microwave Opt. Technol. Lett.*, 48(12):2570–2573, 2006.

[39] D. Chicherin, S. Dudorov, M. Sterner, et al. Micro-fabricated high-impedance surface for millimeter wave beam steering applications. In *2008 33rd International Conference on Infrared, Millimeter and Terahertz Waves*, pages 1–3, 2008.

[40] S. Colburn, A. Zhan, and A. Majumdar. Tunable metasurfaces via subwavelength phase shifters with uniform amplitude. *Sci. Rep.*, 7:40174, 2017.

[41] T. Colomb, E. Cuche, F. Montfort, P. Marquet, and C. Depeursinge. Jones vector imaging by use of digital holography: Simulation and experimentation. *Optics Communications*, 231(1-6):137–147, 2004.

[42] F. Costa, S. Genovesi, A. Monorchio, and G. Manara. A circuit-based model for the interpretation of perfect metamaterial absorbers. *IEEE Transactions on Antennas and Propagation*, 61(3):1201–1209, 2013.

[43] G. Coviello, et al. Effects on phased arrays radiation pattern due to phase error distribution in the phase shifter operation. *MATEC Web of Conferences*, 76:03002, 2016.

[44] T.-J. Cui, S. Liu, and L.-L. Li. Information entropy of coding metasurface. *Light: Science & Applications*, 5(11):e16172, 2016.

[45] T. J. Cui, M. Q. Qi, X. Wan, J. Zhao, and Q. Cheng. Coding metamaterials, digital metamaterials and programmable metamaterials. *Light: Science & Applications*, 3(10):e218, 2014.

[46] N. Dabidian, S. Dutta-Gupta, I. Kholmanov, et al. Experimental demonstration of phase modulation and motion sensing using graphene-integrated metasurfaces. *Nano Lett*, 16:3607–3615, 2016.

[47] A. D'iaz-Rubio, V. S. Asadchy, A. Elsakka, and S. A. Tretyakov. From the generalized reflection law to the realization of perfect anomalous reflectors. *Sci. Adv.*, 3(8):e1602714, 2017.

[48] A. D'iaz-Rubio, J. Li, C. Shen, S. A. Cummer, and S. A. Tretyakov. Power flow-conformal metamirrors for engineering wave reflections. *Science Advances*, 5(2):eaau7288, 2019.

[49] A. D'iaz-Rubio and S. A. Tretyakov. Acoustic metasurfaces for scattering-free anomalous reflection and refraction. *Physical Review B*, 96(12), 2017.

[50] M. Decker, C. Kremers, A. Minovich, et al. Electro-optical switching by liquid-crystal controlled metasurfaces. *Opt. Express*, 21(7):8879–8885, 2013.

[51] F. Dincer. Electromagnetic energy harvesting application based on tunable perfect metamaterial absorber. *J. Electromagn. Waves Appl.*, 29(18):2444–2453, 2015.

[52] T. Driscoll, H.-T. Kim, B.-G. Chae, et al. Memory metamaterials. *Science*, 325(5947):1518–1521, 2009.

[53] T. Driscoll, S. Palit, M. M. Qazilbash, et al. Dynamic tuning of an infrared hybrid-metamaterial resonance using vanadium dioxide. *Appl. Phys. Lett.*, 93(2):024101, 2008.

[54] X. Gan, R.-J. Shiue, Y. Gao, I. Meric, T. F. Heinz, K. Shepard, J. Hone, S. Assefa, and D. Englund. Chip-integrated ultrafast

graphene photodetector with high responsivity. *Nature Photonics*, 7(11):883, 2013.

[55] S. B. Glybovski, S. A. Tretyakov, P. A. Belov, Y. S. Kivshar, and C. R. Simovski. Metasurfaces: From microwaves to visible. *Physics Reports*, 634:1–72, 2016.

[56] N. K. Grady, J. E. Heyes, D. R. Chowdhury, Y. Zeng, M. T. Reiten, A. K. Azad, A. J. Taylor, D. A. R. Dalvit, and H.-T. Chen. Terahertz metamaterials for linear polarization conversion and anomalous refraction. *Science*, 304:1304–1307, 2013.

[57] K. Guan et al. On millimeter wave and THz mobile radio channel for smart rail mobility. *IEEE Transactions on Vehicular Technology*, 66(7):5658–5674, 2017.

[58] J. Gubbi, R. Buyya, S. Marusic, and M. Palaniswami. Internet of things (IoT): A vision, architectural elements, and future directions. *Future Generation Computer Systems*, 29(7):1645–1660, 2013.

[59] K. Gurney. *An Introduction to Neural Networks*. CRC press, 2014.

[60] G. W. Hanson. Dyadic Green's functions for an anisotropic, nonlocal model of biased graphene. *IEEE Transactions on Antennas and Propagation*, 56(3):747–757, 2008.

[61] Y. He, P. He, et al. Role of ferrites in negative index metamaterials. *IEEE J MAG*, 42(10):2852–2854, 2006.

[62] E. Hitzer. Introduction to Clifford's geometric algebra. *arXiv preprint arXiv:1306.1660*, 2013.

[63] C. L. Holloway, et al. An overview of the theory and applications of metasurfaces. *IEEE Ant. and Propag. Mag.*, 54:10–35, 2012.

[64] S. E. Hosseininejad, K. Rouhi, M. Neshat, A. Cabellos-Aparicio, S. Abadal, and E. Alarcon. Digital metasurface based on graphene: An application to beam steering in terahertz plasmonic antennas. *IEEE Transactions on Nanotechnology*, 18(1):734–746, 2019.

[65] S. E. Hosseininejad, K. Rouhi, M. Neshat, R. Faraji-Dana, A. Cabellos-Aparicio, S. Abadal, and E. Alarcon. Reprogrammable graphene-based metasurface mirror with adaptive focal point for THz imaging. *Scientific Reports*, 9:2868, 2019.

[66] H. Hu, F. Zhai, D. Hu, Z. Li, B. Bai, X. Yang, and Q. Dai. Broadly tunable graphene plasmons using an ion-gel top gate with low control voltage. *Nanoscale*, 7(46):19493–19500, 2015.

[67] C. Huang, C. Zhang, J. Yang, et al. Reconfigurable metasurface for multifunctional control of electromagnetic waves. *Adv. Opt. Mater.*, 5(22):1700485, 2017.

[68] Y.-W. Huang, H. W. H. Lee, R. Sokhoyan, R. A. Pala, K. Thyagarajan, S. Han, D. P. Tsai, and H. A. Atwater. Gate-tunable conducting oxide metasurfaces. *Nano Letters*, 16(9):5319–5325, 2016.

[69] P. P. Iyer, M. Pendharkar, and J. A. Schuller. Electrically reconfigurable metasurfaces using heterojunction resonators. *Advanced Optical Materials*, 4(10):1582–1588, 2016.

[70] M. M. Jadidi, A. B. Sushkov, et al. Tunable terahertz hybrid metal–graphene plasmons. *Nano Lett*, 15:7099–7104, 2015.

[71] L. Ju, B. Geng, J. Horng, et al. Graphene plasmonics for tunable terahertz metamaterials. *Nat. Nanotechnol.*, 6:630–634, 2011.

[72] N. Kakenov, O. Balci, T. Takan, V. A. Ozkan, H. Altan, and C. Kocabas. Observation of gate-tunable coherent perfect absorption of terahertz radiation in graphene. *ACS Photonics*, 3(9):1531–1535, 2016.

[73] A. Kannan, N. Enright Jerger, and G. H. Loh. Exploiting interposer technologies to disintegrate and reintegrate multicore processors. *IEEE Micro*, 36(3):84–93, 2016.

[74] Y. G. Karpov. AnyLogic: A new generation professional simulation tool. In *VI International Congress on Mathematical Modeling, Nizni-Novgorog, Russia*, 2004.

[75] G. Kenanakis, R. Zhao, N. Katsarakis, et al. Optically controllable THz chiral metamaterials. *Opt. Express*, 22:12149–12159, 2014.

[76] H. K. Khalil. *Nonlinear Systems*, volume 3. Pearson, 2001.

[77] P. A. Khomyakov, A. A. Starikov, G. Brocks, and P. J. Kelly. Nonlinear screening of charges induced in graphene by metal contacts. *Physical Review B*, 82(11):115437, 2010.

[78] A. V. Kildishev, A. Boltasseva, and V. M. Shalaev. Planar photonics with metasurfaces. *Science*, 339(6125):1232009, 2013.

[79] H. Kim, N. Charipar, E. Breckenfeld, et al. Active terahertz metamaterials based on the phase transition of VO2 thin films. *Thin Solid Films*, 596(Supplement C):45–50, 2015.

[80] H. K. Kim, D. Lee, and S. Lim. Frequency-tunable metamaterial absorber using a varactor-loaded fishnet-like resonator. *Appl. Opt.*, 55(15):4113–4118, 2016.

[81] S. Kim, M. S. Jang, V. W. Brar, K. W. Mauser, L. Kim, and H. A. Atwater. Electronically tunable perfect absorption in graphene. *Nano Letters*, 18(2):971–979, 2018.

[82] A. Komar, Z. Fang, J. Bohn, J. Sautter, M. Decker, A. Miroshnichenko, T. Pertsch, I. Brener, Y. S. Kivshar, I. Staude, and D. N. Neshev. Electrically tunable all-dielectric optical metasurfaces based on liquid crystals. *Appl. Phys. Lett.*, 110(7):071109, 2017.

[83] K. M. Kossifos, L. Petrou, G. Varnava, A. Pitilakis, O. Tsilipakos, F. Liu, P. Karousios, A. Tasolamprou, M. Seckel, D. Manessis, N. V. Kantartzis, D.-H. Kwon, M. A. Antoniades, and J. Georgiou. Toward the realization of a programmable metasurface absorber enabled by custom integrated circuit technology. *arXiv:submit/3093519*, pages 1–12, Submitted Mar. 2020.

[84] P. Kouvaros, D. Kouzapas, A. Philippou, et al. Formal verification of a programmable hypersurface. In *International Workshop on Formal Methods for Industrial Critical Systems*, pages 83–97, 2018.

[85] D. Kouzapas, C. Skitsas, T. Saeed, V. Soteriou, M. Lestas, A. Philippou, S. Abadal, C. Liaskos, L. Petrou, J. Georgiou, and A. Pitsilides. Towards fault adaptive routing in metasurface controller networks. *Journal of Systems Architecture*, 106:101703, 2020.

[86] D.-H. Kwon and S. A. Tretyakov. Perfect reflection control for impenetrable surfaces using surface waves of orthogonal polarization. *Phys. Rev. B*, 96(8):085438, 2017.

[87] S. H. Lee, M. Choi, T.-T. Kim, S. Lee, M. Liu, X. Yin, H. K. Choi, S. S. Lee, C.-G. Choi, S.-Y. Choi, X. Zhang, and B. Min. Switching terahertz waves with gate-controlled active graphene metamaterials. *Nature Materials*, 11(11):936–941, 2012.

[88] P. Levis, S. Madden, J. Polastre, R. Szewczyk, K. Whitehouse, A. Woo, D. Gay, J. Hill, M. Welsh, E. Brewer, et al. TinyOS: An operating system for sensor networks. In *Ambient Intelligence*, pages 115–148. Springer, 2005.

[89] A. Li et al. Metasurfaces and their applications. *Nanophotonics*, 7(6):989–1011, 2018.

[90] L. Li, T. Jun Cui, W. Ji, et al. Electromagnetic reprogrammable coding-metasurface holograms. *Nature Communications*, 8(1):197, 2017.

[91] Y. B. Li, L. L. Li, B. B. Xu, et al. Transmission-type 2-bit programmable metasurface for single-sensor and single-frequency microwave imaging. *Sci. Rep.*, 6:23731, 2016.

[92] C. Liaskos et al. Using any surface to realize a new paradigm for wireless communications. *Communications of the ACM*, 61(11):30–33, 2018.

[93] C. Liaskos, S. Nie, A. Tsioliaridou, A. Pitsillides, S. Ioannidis, and I. Akyildiz. A new wireless communication paradigm through software-controlled metasurfaces. *IEEE Communications Magazine*, 56(9):162–169, 2018.

[94] C. Liaskos, S. Nie, A. Tsioliaridou, A. Pitsillides, S. Ioannidis, and I. Akyildiz. A new wireless communication paradigm through software-controlled metasurfaces. *IEEE Communications Magazine*, 56(9):162–169, 2018.

[95] C. Liaskos, S. Nie, A. Tsioliaridou, A. Pitsillides, S. Ioannidis, and I. Akyildiz. A novel communication paradigm for high capacity and security via programmable indoor wireless environments in next generation wireless systems. *Ad Hoc Networks*, 87:1–16, 2019.

[96] C. Liaskos, S. Nie, A. Tsioliaridou, A. Pitsillides, S. Ioannidis, and I. F. Akyildiz. Mobility-aware beam steering in metasurface-based programmable wireless environments. In *ICASSP'20*, 2020.

[97] C. Liaskos, S. Petridou, and G. Papadimitriou. Cost-Aware Wireless Data Broadcasting. *IEEE Transactions on Broadcasting*, 56(1):66–76, 2010.

[98] C. Liaskos, S. Petridou, and G. Papadimitriou. Towards realizable, low-cost broadcast systems for dynamic environments. *IEEE/ACM Transactions on Networking*, 19(2):383–392, 2010.

[99] C. Liaskos, A. Pitilakis, et al. Initial UML definition of the hypersurface compiler middle-ware. *European Commission Project VISORSURF: Accepted Public Deliverable D2.2, 31-Dec-2017, [Online:] http://www.visorsurf.eu/m/VISORSURF-D2.2.pdf*.

[100] C. Liaskos, A. Tsioliaridou, P. A, G. Pyrialakos, O. Tsilipakos, A. Tasolamprou, N. Kantartzis, S. Ioannidis, M. Kafesaki, A. Pitsillides, and I. F. Akyildiz. Joint compressed sensing and manipulation of wireless emissions with intelligent surfaces. In *IEEE DCOSS 2019, IoT 4.0 Workshop*, pages 1–6, 2019.

[101] C. Liaskos, A. Tsioliaridou, et al. Initial UML definition of the hypersurface programming interface and virtual functions. *European Commission Project VISORSURF: Accepted Public Deliverable D2.1, 31-Dec-2017, [Online:] http://www.visorsurf.eu/m/VISORSURF-D2.1.pdf*.

[102] C. Liaskos, A. Tsioliaridou, S. Nie, A. Pitsillides, S. Ioannidis, and I. F. Akyildiz. An interpretable neural network for configuring programmable wireless environments. In *IEEE SPAWC 2019*, pages 1–5, 2019.

[103] C. Liaskos, A. Tsioliaridou, S. Nie, A. Pitsillides, S. Ioannidis, and I. F. Akyildiz. On the network-layer modeling and configuration of programmable wireless environments. *IEEE/ACM Transactions on Networking*, 27(4):1696–1713, 2019.

[104] C. Liaskos, A. Tsioliaridou, A. Pitsillides, et al. Design and development of software defined metamaterials for nanonetworks. *IEEE M CAS*, 15(4):12–25, 2015.

[105] C. Liaskos, A. Xeros, G. Papadimitriou, M. Lestas, and A. Pitsillides. Broadcast scheduling with multiple concurrent costs. *IEEE Transactions on Broadcasting*, 58(2):178–186, 2012.

[106] F. Liu, A. Pitilakis, M. S. Mirmoosa, et al. Programmable metasurfaces: State of the art and prospects. In *Proceedings of the ISCAS '18*, 2018.

[107] F. Liu, A. Pitilakis, M. S. Mirmoosa, O. Tsilipakos, X. Wang, A. C. Tasolamprou, S. Abadal, A. Cabellos-Aparicio, E. Alarcon, C. Liaskos, N. V. Kantartzis, M. Kafesaki, E. N. Economou, C. M. Soukoulis, and S. Tretyakov. Programmable metasurfaces: State of the art and prospects. In *Proceedings of the ISCAS '18*, 2018.

[108] F. Liu, O. Tsilipakos, A. Pitilakis, A. C. Tasolamprou, M. S. Mirmoosa, N. V. Kantartzis, D.-H. Kwon, J. Georgiou, K. Kossifos, M. A. Antoniades, M. Kafesaki, C. M. Soukoulis, and S. A. Tretyakov. Intelligent metasurfaces with continuously tunable local surface impedance for multiple reconfigurable functions. *Phys. Rev. Appl.*, 11(04):044024, 2019.

[109] F. Liu, O. Tsilipakos, A. Pitilakis, A. C. Tasolamprou, M. S. Mirmoosa, N. V. Kantartzis, D.-H. Kwon, M. Kafesaki, C. M. Soukoulis, and S. A. Tretyakov. Intelligent metasurfaces with continuously tunable local surface impedance for multiple reconfigurable functions. *Physical Review Applied*, 11(4), 2019.

[110] S. Liu, T. J. Cui, A. Noor, Z. Tao, H. C. Zhang, G. D. Bai, Y. Yang, and X. Y. Zhou. Negative reflection and negative surface wave conversion from obliquely incident electromagnetic waves. *Light: Science and Applications*, 7(5):18008–18011, 2018.

[111] S. Liu, Y. Long, et al. Bioinspired adaptive microplate arrays for magnetically tuned optics. *Adv. Opt. Mater.*, 5:1601043, 2017.

[112] Z. Luo, J. Long, X. Chen, and D. Sievenpiper. Electrically tunable metasurface absorber based on dissipating behavior of embedded varactors. *Appl. Phys. Lett.*, 109(7):071107, 2016.

[113] O. Luukkonen, C. R. Simovski, G. Granet, G. Goussetis, D. Lioubtchenko, A. V. Raisanen, and S. A. Tretyakov. Simple and accurate analytical model of planar grids and high-impedance surfaces comprising metal strips or patches. *IEEE*

Transactions on Antennas and Propagation, 56(6):1624–1632, 2008.

[114] B. Ma, S. Liu, X. Kong, et al. A novel wide-band tunable metamaterial absorber based on varactor diode/graphene. *Optik - International Journal for Light and Electron Optics*, 127:3039–3043, 2016.

[115] M. M. Masud, B. Ijaz, I. Ullah, and B. Braaten. A compact dual-band EMI metasurface shield with an actively tunable polarized lower band. *IEEE J EMC*, 54(5):1182–1185, 2012.

[116] N. F. Maxemchuk. Regular mesh topologies in local and metropolitan area networks. *AT&T Technical Journal*, 64(7):1659–1685, 1985.

[117] C. Mias and J. H. Yap. A varactor-tunable high impedance surface with a resistive-lumped-element biasing grid. *IEEE Trans. Antennas Propag.*, 55(7):1955–1962, 2007.

[118] R. Moore. *Interval Analysis*. Prentice Hall, 1966.

[119] D. Morits, M. Morits, V. Ovchinnikov, M. Omelyanovich, A. Tamminen, S. Tretyakov, and C. Simovski. Multifunctional stretchable metasurface for the THz range. *J. Opt.*, 16(3):032001, 2014.

[120] K. S. Novoselov, A. K. Geim, S. Morozov, D. Jiang, M. Katsnelson, I. Grigorieva, S. Dubonos, and AA. Firsov. Two-dimensional gas of massless Dirac fermions in graphene. *Nature*, 438(7065):197, 2005.

[121] J.-Y. Ou, E. Plum, J. Zhang, and N. I. Zheludev. An electromechanically reconfigurable plasmonic metamaterial operating in the near-infrared. *Nat. Nanotechnol.*, 8(4):252–255, 2013.

[122] W. J. Padilla, A. J. Taylor, C. Highstrete, et al. Dynamical electric and magnetic metamaterial response at terahertz frequencies. *Phys. Rev. Lett.*, 96(10):107401, 2006.

[123] R. K. Pandey, W. A. Stapleton, P. Padmini, et al. Magnetically tuned varistor-transistor hybrid device. *AIP Adv*, 2:042188, 2016.

[124] Y.-H. Pao and V. Varatharajulu. Huygens' principle, radiation conditions, and integral formulas for the scattering of elastic

waves. *The Journal of the Acoustical Society of America*, 59(6), 1976.

[125] G. Pawlik, K. Tarnowski, W. Walasik, et al. Infrared cylindrical cloak in nanosphere dispersed liquid crystal metamaterial. *Opt. Lett.*, 37(11):1847–1849, 2012.

[126] N. Perez-Arancibia, J. Gibson, and T. Tsao. Laser beam pointing and stabilization by intensity feedback control. In *American Control Conference, ACC 2009*, pages 2837–2842, 2009.

[127] L. Petrou and J. Georgiou. Hardware design and manufacturing of the hypersurface control nodes and hardware interface, UCY internal report, July 2017.

[128] L. Petrou, P. Karousios, and J. Georgiou. Asynchronous circuits as an enabler of scalable and programmable metasurfaces. In *Proceedings of the ISCAS '18*, 2018.

[129] E. Pilakoutas. Development, implementation, and experimental evaluation of routing algorithms in a Manhattan network topology [report]. 2018.

[130] A. Pitilakis, A. C. Tasolamprou, C. Liaskos, F. Liu, O. Tsilipakos, X. Wang, M. S. Mirmoosa, K. Kossifos, J. Georgiou, A. Pitsilides, N. V. Kantartzis, S. Ioannidis, E. N. Economou, M. Kafesaki, S. A. Tretyakov, and C. M. Soukoulis. Software-defined metasurface paradigm: Concept, challenges, prospects. In *Proceedings of the METAMATERIALS '18*. IEEE, 2018.

[131] A. Pitilakis, O. Tsilipakos, F. Liu, K. M. Kossifos, A. C. Tasolamprou, D.-H. Kwon, M. S. Mirmoosa, D. Manessis, N. V. Kantartzis, C. Liaskos, M. A. Antoniades, J. Georgiou, C. M. Soukoulis, M. Kafesaki, and S. A. Tretyakov. A multi-functional intelligent metasurface: Electromagnetic design accounting for fabrication aspects, arXiv:2003.08654, 2020.

[132] The VISORSURF project. A hardware platform for software-driven functional metasurfaces. *Horizon 2020 Future and Emerging Technologies*, 2017.

[133] O. Rabinovich and A. Epstein. Analytical design of printed circuit board (PCB) metagratings for perfect anomalous reflection. *IEEE Trans. Antennas Propag.*, 66(8):4086–4095, 2018.

[134] Y. Ra'Di, C. R. Simovski, and S. A. Tretyakov. Thin perfect absorbers for electromagnetic waves: Theory, design, and realizations. *Phys. Rev. Appl*, 3(3):037001, 2015.

[135] Y. Ra'di, D. L. Sounas, and A. Alu. Metagratings: Beyond the limits of graded metasurfaces for wave front control. *Phys. Rev. Lett.*, 119(6):067404, 2017.

[136] W. Roh, J.-Y. Seol, J. Park, B. Lee, J. Lee, Y. Kim, J. Cho, K. Cheun, and F. Aryanfar. Millimeter-wave beamforming as an enabling technology for 5G cellular communications: Theoretical feasibility and prototype results. *IEEE Communications Magazine*, 52(2):106–113, 2014.

[137] T. Saeed, S. Abadal, C. Liaskos, A. Pitsillides, H. Taghvaee, A. Cabellos-Aparicio, M. Lestas, and E. Alarcon. Workload characterization of programmable metasurfaces. In *Proceeding of the NANOCOM '19*, 2019.

[138] T. Saeed, C. Skitsas, D. Kouzapas, M. Lestas, V. Soteriou, A. Philippou, S. Abadal, C. Liaskos, L. Petrou, J. Georgiou, et al. Fault adaptive routing in metasurface controller networks. In *2018 11th International Workshop on Network on Chip Architectures (NoCArc)*, pages 1–6, 2018.

[139] T. Saeed, C. Skitsas, D. Kouzapas, M. Lestas, V. Soteriou, A. Philippou, S. Abadal, C. Liaskos, L. Petrou, J. Georgiou, et al. Fault adaptive routing in metasurface controller networks. In *2018 11th International Workshop on Network on Chip Architectures (NoCArc)*, pages 1–6, 2018.

[140] T. Saeed, C. Skitsas, D. Kouzapas, M. Lestas, V. Soteriou, A. Philippou, S. Abadal, C. Liaskos, L. Petrou, J. Georgiou, and A. Pitsillides. Fault adaptive routing in metasurface network controllers. In *Proceedings of the NoCArc '18*, 2018.

[141] A. Safaei, S. Chandra, A. Vazquez-Guardado, J. Calderon, D. Franklin, L. Tetard, L. Zhai, M. N. Leuenberger, and D. Chanda. Dynamically tunable extraordinary light absorption in monolayer graphene. *Physical Review B*, 96(16):165431, 2017.

[142] B. Sensale-Rodriguez, R. Yan, M. M. Kelly, T. Fang, K. Tahy, W. S. Hwang, D. Jena, L. Liu, and H. G. Xing. Broadband graphene terahertz modulators enabled by intraband transitions. *Nature Communications*, 3(1):1–7, 2012.

[143] M. Seo, J. Kyoung, H. Park, et al. Active terahertz nanoantennas based on VO2 phase transition. *Nano Lett*, 10:2064–2068, 2010.

[144] I. V. Shadrivov, P. V. Kapitanova, S. I. Maslovski, and Y. S. Kivshar. Metamaterials controlled with light. *Physical Review Letters*, 109(8):083902, 2012.

[145] N.-H. Shen, M. Massaouti, M. Gokkavas, et al. Optically implemented broadband blueshift switch in the terahertz regime. *Phys. Rev. Lett.*, 106(3):037403, 2011.

[146] J.-H. Shin, K. Moon, E. S. Lee, et al. Metal-VO2 hybrid grating structure for a terahertz active switchable linear polarizer. *Nanotechnology*, 26(31):315203, 2015.

[147] D. Shrekenhamer, W.-C. Chen, and W. J. Padilla. Liquid crystal tunable metamaterial absorber. *Phys. Rev. Lett.*, 110:177403, 2013.

[148] B. Sieklucka and D. Pinkowicz. *Molecular Magnetic Materials: Concepts and Applications*. John Wiley & Sons, 2016.

[149] D. F. Sievenpiper, J. H. Schaffner, H. J. Song, et al. Two-dimensional beam steering using an electrically tunable impedance surface. *IEEE Trans. Antennas Propag.*, 51(10):2713–2722, 2003.

[150] K. J. Singh and D. S. Kapoor. Create your own Internet of Things: A survey of IoT platforms. *IEEE Consumer Electronics Magazine*, 6(2):57–68, 2017.

[151] J. Su, Y. Lu, H. Zhang, Z. Li, Y. Lamar Yang, Y. Che, and K. Qi. Ultra-wideband, wide angle and polarization-insensitive specular reflection reduction by metasurface based on parameter-adjustable meta-atoms. *Scientific Reports*, 7:42283, 2017.

[152] L. Subrt and P. Pechac. Intelligent walls as autonomous parts of smart indoor environments. *IET Communications*, 6(8), 2012.

[153] C. Sun, C.-H. O. Chen, G. Kurian, L. Wei, J. Miller, A. Agarwal, L.-S. Peh, and V. Stojanovic. DSENT-A tool connecting emerging photonics with electronics for opto-electronic networks-on-chip modeling. In *2012 IEEE/ACM Sixth International Symposium on Networks-on-Chip*, pages 201–210. IEEE, 2012.

[154] H. Taghvaee, S. Abadal, J. Georgiou, A. Cabellos-Aparicio, and E. Alarcon. Fault tolerance in programmable metasurfaces: The beam steering case. In *2019 IEEE International Symposium on Circuits and Systems, ISCAS 2019*, pages 1–5. IEEE Computer Society, 2019.

[155] H. Taghvaee, A. Cabellos-Aparicio, J. Georgiou, and S. Abadal. Error analysis of programmable metasurfaces for beam steering. *IEEE Journal on Emerging and Selected Topics in Circuits and Systems*, 10(1):62–74, 2020.

[156] X. Tan et al. Increasing indoor spectrum sharing capacity using smart reflect-array. In *IEEE ICC'16*.

[157] X. Tan, Z. Sun, D. Koutsonikolas, and J. M. Jornet. Enabling indoor mobile millimeter-wave networks based on smart reflect-arrays. In *Proceedings of the INFOCOM '18*, pages 270–278, 2018.

[158] Y.-W. Tan, Y. Zhang, K. Bolotin, Y. Zhao, S. Adam, E. H. Hwang, S. D. Sarma, H. L. Stormer, and P. Kim. Measurement of scattering rate and minimum conductivity in graphene. *Physical Review Letters*, 99(24):246803, 2007.

[159] Z. Tao, X. Wan, B. C. Pan, and T. J. Cui. Reconfigurable conversions of reflection, transmission, and polarization states using active metasurface. *Appl. Phys. Lett.*, 110(12):121901, 2017.

[160] A. C. Tasolamprou, A. D. Koulouklidis, C. Daskalaki, C. P. Mavidis, G. Kenanakis, G. Deligeorgis, Z. Viskadourakis, P. Kuzhir, S. Tzortzakis, M. Kafesaki, et al. Experimental demonstration of ultrafast thz modulation in a graphene-based thin film absorber through negative photoinduced conductivity. *ACS Photonics*, 6(3):720–727, 2019.

[161] A. C. Tasolamprou, M. S. Mirmoosa, O. Tsilipakos, A. Pitilakis, F. Liu, S. Abadal, A. Cabellos-Aparicio, E. Alarcon, C. Liaskos, N. V. Kantartzis, S. Tretyakov, M. Kafesaki, E. N. Economou, and C. M. Soukoulis. Intercell wireless communication in software-defined metasurfaces. In *Proceedings of the ISCAS '18*, 2018.

[162] A. C. Tasolamprou, A. Pitilakis, S. Abadal, O. Tsilipakos, X. Timoneda, H. Taghvaee, M. S. Mirmoosa, F. Liu, C. Liaskos,

A. Tsioliaridou, S. Ioannidis, N. V. Kantartzis, D. Manessis, J. Georgiou, A. Cabellos-Aparicio, E. Alarcon, A. Pitsillides, I. F. Akyildiz, S. A. Tretyakov, E. N. Economou, M. Kafesaki, and C. M. Soukoulis. Exploration of intercell wireless millimeter-wave communication in the landscape of intelligent metasurfaces. *IEEE Access*, 7:122931–122948, 2019.

[163] H. Tataria, F. Tufvesson, and O. Edfors. Real-time implementation aspects of large intelligent surfaces. *arXiv preprint arXiv:2003.01672*, 2020.

[164] X. J. Technologies. *The AnyLogic Simulator*. 2018.

[165] S. Thongrattanasiri, F. H. L. Koppens, and F. J. G. de Abajo. Complete optical absorption in periodically patterned graphene. *Physical Review Letters*, 108(4):047401, 2012.

[166] S. Tretyakov. *Analytical Modeling in Applied Electromagnetics*. Artech House, 2003.

[167] S. A. Tretyakov. Metasurfaces for general transformations of electromagnetic fields. *Philosophical Transactions of the Royal Society A: Mathematical, Physical and Engineering Sciences*, 373(2049):20140362, 2015.

[168] S. A. Tretyakov and C. R. Simovski. Dynamic model of artificial reactive impedance surfaces. *Journal of Electromagnetic Waves and Applications*, 17(1):131–145, 2003.

[169] O. Tsilipakos, A. C. Tasolamprou, T. Koschny, et al. Pairing toroidal and magnetic dipole resonances in elliptic dielectric rod metasurfaces for reconfigurable wavefront manipulation in reflection. *Advanced Optical Materials*, 6(22), 2018.

[170] A. Tsioliaridou, C. Liaskos, A. Pitsillides, and S. Ioannidis. A novel protocol for network-controlled metasurfaces. In *ACM NANOCOM'17*, pages 3:1–3:6, Sept. 2019.

[171] I. T. Union. Report of the CPM to WRC-19. In *World Radiocommunication Conference*, 2019.

[172] A. Vakil and N. Engheta. Transformation optics using graphene. *Science*, 332(6035):1291–1294, 2011.

[173] X. Vilajosana, P. Tuset, T. Watteyne, and K. Pister. Openmote: Open-source prototyping platform for the industrial IoT. In *International Conference on Ad Hoc Networks*, pages 211–222. Springer, 2015.

[174] H. Wakatsuchi et al. Waveform-dependent absorbing metasurfaces. *Physical Review Letters*, 111(24), 2013.

[175] D. Wang, L. Zhang, Y. Gong, et al. Multiband switchable terahertz quarter-wave plates via phase-change metasurfaces. *IEEE Photonics Journal*, 8(1), 2016.

[176] D. Wang, L. Zhang, Y. Gu, et al. Switchable ultrathin quarter-wave plate in terahertz using active phase-change metasurface. *Sci. Rep.*, 5(15020), 2015.

[177] Q. Wang, E. T. F. Rogers, B. Gholipour, C.-M. Wang, G. Yuan, J. Teng, and N. I. Zheludev. Optically reconfigurable metasurfaces and photonic devices based on phase change materials. *Nat. Photonics*, 10(1):60–65, 2015.

[178] X.-C. Wang and S. A. Tretyakov. Tunable perfect absorption in continuous graphene sheets on metasurface substrates. *arXiv preprint:1712.01708*, 2017.

[179] A. Welkie et al. Programmable radio environments for smart spaces. In *ACM HOTNETS'17*, pages 36–42, 2017.

[180] B. Widrow, P. E. Mantey, L. J. Griffiths, and B. B. Goode. Adaptive antenna systems. *Proceedings of the IEEE*, 55(12):2143–2159, 1967.

[181] A. M. H. Wong and G. V. Eleftheriades. Perfect anomalous reflection with a bipartite huygens' metasurface. *Phys. Rev. X*, 8(1):011036, 2018.

[182] Q. Wu and R. Zhang. Intelligent reflecting surface enhanced wireless network. *arXiv:1809.01423*, 2018.

[183] R. Y. Wu, C. B. Shi, S. Liu, W. Wu, and T. J. Cui. Addition theorem for digital coding metamaterials. *Advanced Optical Materials*, 6(5):1–10, 2018.

[184] F. Xia, T. Mueller, Y.-m. Lin, A. Valdes-Garcia, and P. Avouris. Ultrafast graphene photodetector. *Nature Nanotechnology*, 4(12):839, 2009.

[185] H.-X. Xu, S. Tang, S. Ma, et al. Tunable microwave metasurfaces for high-performance operations: Dispersion compensation and dynamical switch. *Sci. Rep.*, 6:38255, 2016.

[186] A. Yakovlev, M. G. Silveirinha, O. Luukkonen, C. R. Simovski, I. S. Nefedov, and S. A. Tretyakov. Characterization of surface-wave and leaky-wave propagation on wire-medium slabs and mushroom structures based on local and nonlocal homogenization models. *IEEE Transactions on Microwave Theory and Techniques*, 57(11):2700–2714, 2009.

[187] H. Yang, X. Cao, F. Yang, J. Gao, S. Xu, and M. Li. A programmable metasurface with dynamic polarization, scattering and focusing control. *Scientific Reports*, (October):1–11, 2016.

[188] H. Yang, X. Cao, F. Yang, J. Gao, S. Xu, M. Li, X. Chen, Y. Zhao, Y. Zheng, and S. Li. A programmable metasurface with dynamic polarization, scattering and focusing control. *Scientific Reports*, 6(35692), 2016.

[189] Y. Yao, M. A. Kats, R. Shankar, Y. Song, J. Kong, M. Loncar, and F. Capasso. Wide wavelength tuning of optical antennas on graphene with nanosecond response time. *Nano Letters*, 14(1):214–219, 2013.

[190] X. Ying, Y. Pu, Y. Luo, H. Peng, Z. Li, Y. Jiang, J. Xu, and Z. Liu. Enhanced universal absorption of graphene in a Salisbury screen. *Journal of Applied Physics*, 121(2):023110, 2017.

[191] Y. Yong-Jun, H. Yong-Jun, W. Guang-Jun, et al. Tunable broadband metamaterial absorber consisting of ferrite slabs and a copper wire. *Chin. Phys. B*, 21(3):038501, 2012.

[192] N. Yu and F. Capasso. Flat optics with designer metasurfaces. *Nature Materials*, 13(2):139–150, 2014.

[193] N. Yu, P. Genevet, M. A. Kats, F. Aieta, J.-P. Tetienne, F. Capasso, and Z. Gaburro. Light propagation with phase discontinuities: Generalized laws of reflection and refraction. *Science*, 334(October):333–337, 2011.

[194] A. Zappone, M. Di Renzo, F. Shams, X. Qian, and M. Debbah. Overhead-aware design of reconfigurable intelligent surfaces in smart radio environments. *arXiv preprint arXiv:2003.02538*, 2020.

[195] F. Zhang, Q. Zhao, W. Zhang, et al. Voltage tunable short wire-pair type of metamaterial infiltrated by nematic liquid crystal. *Appl. Phys. Lett.*, 97(13):134103, 2010.

[196] L. Zhang, X. Q. Chen, S. Liu, Q. Zhang, J. Zhao, J. Y. Dai, G. D. Bai, X. Wan, Q. Cheng, G. Castaldi, V. Galdi, and T. J. Cui. Space-time-coding digital metasurfaces. *Nature Communications*, 9(1):4334, 2018.

[197] L. Zhang, S. Liu, L. Li, and T. J. Cui. Spin-controlled multiple pencil beams and vortex beams with different polarizations generated by Pancharatnam-Berry coding metasurfaces. *ACS Applied Materials and Interfaces*, 9(41):36447–36455, 2017.

[198] S. Zhang, J. Zhou, Y.-S. Park, et al. Photoinduced handedness switching in terahertz chiral metamolecules. *Nat. Commun.*, 3:942, 2012.

[199] J. Zhao, Q. Cheng, J. Chen, et al. A tunable metamaterial absorber using varactor diodes. *New J. Phys.*, 15:043049, 2013.

[200] X. Zhou, Z. Zhang, Y. Zhu, Y. Li, S. Kumar, A. Vahdat, B. Y. Zhao, and H. Zheng. Mirror mirror on the ceiling: Flexible wireless links for data centers. *ACM SIGCOMM Computer Communication Review*, 42(4):443–454, 2012.

[201] Y. Zhou, G. Zhang, H. Chen, P. Zhou, X. Wang, L. Zhang, and L. Zhang. Design of phase gradient coding metasurfaces for broadband wave modulating. *Scientific Reports*, 8:8672, 2018.

[202] B. Zhu, Y. Feng, J. Zhao, et al. Switchable metamaterial reflector/absorber for different polarized electromagnetic waves. *Appl. Phys. Lett.*, 97(5):051906, 2010.

[203] J. Zhu, D. Li, S. Yan, et al. Tunable microwave metamaterial absorbers using varactor-loaded split loops. *EPL*, 112(5):54002, 2015.

[204] D. C. Zografopoulos and R. Beccherelli. Tunable terahertz fishnet metamaterials based on thin nematic liquid crystal layers for fast switching. *Sci. Rep.*, 5:13137, 2015.

Index

ABCD-parameters, 45
absorber, 8, 9, 18, 22, 25, 48, 50, 61,
    67, 68, 70, 72, 73, 325, 328,
    331, 332, 335, 336, 339, 340
actuation, 140, 199, 201, 220, 246
amplitude, 12, 13, 20, 26, 28, 31, 37,
    38, 40, 45, 46, 52, 53, 55,
    109, 119, 120, 127, 231, 324
angle, 19, 20, 22, 23, 33, 50, 52, 54,
    58, 60, 61, 73, 74, 78, 106,
    110, 111, 115, 118, 123, 124,
    130, 138, 152, 153, 189, 193,
    234, 236, 238, 239, 241, 242,
    251, 258, 259, 262, 267, 296,
    302, 307, 321, 335, 341
anisotropic, 326
antenna, 7, 18, 20, 57, 124, 137, 195,
    212, 271, 274, 276, 284, 294,
    298, 306, 308, 320, 323, 338,
    341
aperture, 243, 248, 299, 308
API, 31, 78, 88, 96, 103, 104, 131,
    132, 134, 138, 140, 165, 176
architecture, 32, 45, 47, 77, 79, 143,
    192, 205, 206, 220, 222, 229,
    266, 319, 320, 324, 328
asynchronous, 144, 155, 156, 161,
    164, 165, 168, 170, 177, 207,
    216, 255, 333
attenuation, 96, 278, 280, 308
automata, 156, 323

backplane, 46, 47, 49, 51, 56, 113,
    234
back-propagation, 270, 294, 296
bandwidth, 50, 51, 104, 221, 246, 320
beam, 25, 27, 29, 30, 58, 59, 61, 89,
    104, 117, 118, 121, 127, 128,
    137, 152, 188, 189, 193, 195,
    197, 228, 230, 237, 239, 242,
    248, 262, 299, 307, 308, 323,
    324, 326, 330, 333, 335, 336
beam-steering, 192
bidirectional, 145, 274, 283
Bloch-Floquet, 47, 341
broadband, 27, 334, 335, 339, 340

callback, 79, 133, 164, 167, 176
capacitance, 16, 17, 29, 33, 41, 42,
    44, 46, 47, 50, 54, 60, 65,
    70, 72, 80, 93, 94, 121, 125,
    234, 236, 245
carrier, 18, 27, 28, 62, 63, 65, 209,
    211
cells, 10, 17, 20, 22, 30, 31, 34, 35,
    37, 40, 42, 44, 48, 53, 54,
    57, 66, 89, 94, 117, 119,
    120, 125, 127, 130, 134, 136,
    137, 153, 189, 193, 195, 228,
    230, 236, 241, 254, 266, 267
characterization, 8, 57, 142, 152, 228,
    257, 267, 334, 339
coding-metasurface, 28, 329
coefficient, 11, 45, 48, 55, 56, 65,
    71, 109, 115, 125, 234, 239,
    341
collimation, 121, 273, 275, 278, 279,
    281, 293, 296, 299, 303, 308
communication, 3, 4, 31, 79, 80, 88,
    97, 142, 148, 155, 159, 164,
    165, 167, 173, 176, 184, 188,
    196, 197, 199, 200, 205, 210,
    212, 215, 220, 223, 225, 229,
    252, 255, 267, 270, 274, 284,
    286, 288, 289, 297, 308, 320,
    321, 329, 336, 337, 340

compiler, 31, 77, 78, 91, 97, 99, 104, 130, 132, 135, 138, 140, 217, 330

complexity, 4, 28, 30, 81, 203, 212, 229, 230, 232, 242, 245, 262, 320

components, 6, 9, 24, 32, 88, 107, 111, 121, 127, 142, 144, 145, 151, 169, 190, 203, 207, 213, 216, 217, 225, 226, 246, 247, 283, 287, 305, 309

conductance, 43, 62

conjugate, 128

controller, 31, 78, 83, 85, 99, 121, 141, 153, 158, 165, 170, 176, 188, 199, 200, 202, 203, 205, 206, 208, 211, 212, 215, 223, 228, 245, 252, 254, 257, 259, 264, 297, 328, 334

co-polarized, 122

crystal-based, 25

database, 77, 80, 84, 85, 88, 90, 91, 94, 99, 104, 134, 135, 161, 164, 168, 173, 176, 222

deadlock, 148, 154, 158

Delay-sensitive, 188

dielectric, 17, 33, 43, 44, 49, 51, 62, 68, 74, 101, 246, 337

dimensions, 32, 42, 43, 51, 67, 71, 108, 111, 114, 118, 190, 231, 234, 242, 264, 298

dipole-dipole, 27

directionality, 249

Doppler, 4, 270, 271, 289, 292, 302, 303, 309, 311, 320

eavesdropping, 4, 269, 271, 289, 290, 292, 293, 309, 311

emulator, 142, 159, 176, 177, 186, 197

excitation, 11, 20, 40, 61, 105, 116

fault-tolerance, 151

feed-forward, 270, 294, 307, 309, 316

Fermi-level, 70

Field-Programmable, 30, 341

Floquet, 47, 78, 106, 114, 123, 125, 127, 132, 232, 233, 341

flow-conformal, 325

focal, 23, 29, 128, 273, 276, 278, 279, 327

functionality, 8, 21, 28, 32, 48, 58, 71, 78, 79, 82, 84, 97, 99, 101, 111, 124, 125, 127, 129, 130, 142, 144, 149, 152, 158, 160, 165, 167, 169, 189, 207, 212, 225, 229, 230, 233, 237, 238, 246, 250, 279, 280, 282, 283, 286, 298, 308

gains, 203, 288, 294, 297, 304, 309

gateway, 31, 32, 79, 83, 85, 142, 143, 146, 149, 153, 155, 158, 160, 161, 164, 165, 167, 169, 178, 186, 187, 190, 191, 196, 197, 200, 208, 228, 245, 246, 248, 252, 254

graphene, 9, 26, 62, 323, 326, 332, 334, 336, 341

gyroscopes, 299

half-wave, 114

handedness, 27, 340

handshaking, 143

huygens, 30, 236, 271, 324, 332, 338

hypersurface, 31, 51, 78, 88, 89, 91, 93, 96, 101, 103, 129, 131, 132, 138, 148, 154, 159, 163, 168, 174, 175, 178, 188, 200, 201, 206, 212, 223, 227, 244, 246, 249, 252, 269, 272, 273, 277, 278, 282, 294, 295, 304, 308, 309, 319, 320, 328, 330, 333

identification, 135, 154, 156, 164, 167, 170, 172

illumination, 19, 22, 73, 95, 96, 137, 255, 256

impedance, 7, 9, 12, 22, 24, 26, 29, 37, 42, 63, 67, 69, 80, 93,

94, 102, 104, 109, 223, 277, 278, 331, 332, 335, 337, 341

impinging, 3, 34, 44, 45, 79, 83, 87, 89, 95, 96, 100, 109, 111, 117, 122, 123, 127, 270, 272, 276, 280, 283, 294, 299, 305, 307

incidence, 9, 13, 19, 20, 22, 23, 32, 40, 42, 46, 51, 57, 64, 68, 74, 99, 106, 107, 110, 118, 121, 124, 127, 130, 152, 153, 189, 231, 232, 237, 238, 241, 254, 255, 264, 341

interface, 31, 57, 60, 77, 78, 84, 97, 138, 140, 142, 156, 160, 164, 165, 168, 172, 173, 176, 177, 179, 199, 202, 206, 207, 212, 216, 217, 222, 298, 330, 333

interference, 4, 21, 83, 143, 181, 208, 209, 277, 288, 289, 293, 301, 305, 320, 321

Internet-of-materials, 77, 79, 84, 319

Internet-of-Things, 319

interpolation, 83

interpretable, 270, 297, 308, 309, 330

interrupt, 84, 88, 97, 168, 169

isotropic, 48, 55, 119, 129

Jacobian, 194

Java, 154, 155, 297

KShortestPaths, 290, 310

laminate, 33, 49

layer, 9, 10, 18, 19, 32, 42, 49, 62, 64, 66, 67, 70, 73, 81, 143, 160, 201, 207, 208, 219, 220, 225, 226, 247, 271, 290, 296, 297, 307

lens, 321

liquid-crystal, 325

lumped-element, 9, 341

Lyapunov, 193

magnetic, 8, 10, 12, 18, 24, 36, 40, 41, 74, 332, 335, 337

Magnetic-sensitive, 25

magnitude, 113, 114, 117, 125, 136, 240, 241, 266

manipulation, 30, 53, 55, 100, 101, 235, 237, 245, 272, 330, 337

MEMS, 29, 341

meta-atom, 25, 102, 143, 279, 299, 319

metal-graphene, 72

metallization, 33, 49

metamaterial, 226, 322, 325, 328, 332, 333, 335, 339, 340

metasurface, 4, 9, 10, 12, 18, 22, 24, 28, 37, 42, 46, 55, 57, 63, 67, 68, 70, 77, 79, 81, 97, 99, 108, 114, 116, 123, 127, 134, 135, 143, 145, 152, 153, 190, 202, 225, 228, 230, 241, 246, 249, 250, 252, 254, 261, 266, 280, 298, 319, 323, 325, 331, 338, 339

microelectromechanical, 29, 341

microwave, 26, 27, 31, 45, 49, 62, 201, 321, 323, 324, 329, 339, 340

middleware, 79

mmWave, 195

nanoantennas, 335

nanocontroller, 31

NaNoNetworking, 227

Nanophotonics, 329

nanostructures, 27

near-field, 20, 128, 324

network, 6, 30, 37, 45, 49, 78, 79, 83, 85, 109, 138, 141, 148, 149, 152, 159, 169, 174, 175, 177, 179, 186, 196, 200, 206, 207, 209, 215, 219, 223, 225, 228, 229, 249, 254, 255, 257, 259, 260, 264, 270, 271, 294, 295, 297, 307, 316, 319, 320, 330, 333, 334, 338

network-on-chip, 143, 324

optimization, 4, 38, 51, 52, 54,
78, 89, 99, 104, 105, 115,
124, 125, 130, 134, 140, 269,
287, 288, 293, 298, 319, 320
opto-electronic, 335

ParameterizedFunction, 97
particle, 11, 16, 17
patch-array, 70
phase-change, 338
phased-array, 23
phase-gradient, 38
Photoconductive, 27
photoexcitation, 27, 341
photoinduced, 336, 340
photonics, 327, 335, 336, 338
piezoelectric, 29
plasmonics, 327
polarization, 8, 9, 11, 14, 15, 19, 28,
34, 36, 37, 40, 44, 45, 48,
52, 60, 75, 78, 79, 87, 89,
91, 93, 95, 99, 107, 109,
111, 118, 119, 121, 127,
128, 130, 136, 137, 188,
271, 273, 275, 279, 280,
326, 329, 336, 339
programmable, 30, 152, 202, 225,
226, 229, 237, 250, 269, 270,
277, 278, 282, 283, 286, 308,
309, 319, 320, 322, 325, 328,
333, 334, 336, 338, 339
propagation, 3, 9, 19, 23, 107, 109,
208, 269, 270, 272, 276, 283,
292, 294, 300, 301, 305, 313,
317, 322, 324, 326, 332, 339
protocol, 6, 79, 138, 142, 148, 159,
164, 165, 168, 176, 177, 179,
184, 196, 206, 210, 217, 222,
223, 337
PWE, 31, 270, 271, 274, 287, 289,
290, 294, 297, 304, 308, 309,
312

quasi-continuous, 30, 54
quasistatic, 19, 341

radiation-pattern, 106
random-access, 217
reactance, 16, 18, 33, 34, 43, 50, 57,
64, 66, 70, 93, 116, 121
reciprocity, 38, 40, 128
reflectance, 65
reflectarray, 323
resistance, 13, 15, 16, 18, 22, 33, 41,
46, 47, 54, 55, 57, 60, 64, 70,
72, 80, 93, 94, 125, 202, 203
resistive-lumped-element, 332
retro-reflection, 55, 117

scalability, 140, 143, 223, 226, 232,
239, 251, 321
scattering, 4, 17, 29, 45, 48, 53, 54,
62, 78, 102, 105, 106, 109,
118, 127, 137, 139, 233, 236,
237, 282, 298, 321, 332, 336,
339
SDM, 32, 306
semiconductor, 18, 27, 201, 341
ShortestPath, 290, 291, 312
signal-to-interference, 271, 309
simulator, 130, 142, 151, 154, 257,
264, 337
software-defined, 5, 8, 25, 31, 75,
322, 333, 336, 341
source, 12, 57, 67, 84, 89, 91, 96, 98,
118, 128, 135, 151, 192, 213,
214, 216, 217, 219, 222, 247,
250, 255, 256, 274, 321
Space-time-coding, 340
spectrum, 29, 70, 73, 100, 111, 112,
115, 208, 211, 320, 336
spherical-wave, 19
super-cell, 99, 106, 110, 111, 114, 127
synchronization, 177, 186
synchronous, 155, 156, 206, 207
system-on-a-chip, 214

Telecommunications, 323
terahertz, 26, 62, 65, 67, 69, 71, 74,
248, 322, 324, 326, 332, 334,
335, 338, 340, 341

testbed, 105, 134, 138, 143

Thermal-sensitive, 25

timeout, 164, 167, 170, 181, 183,
  197

tolerance, 115, 116, 124, 151, 154,
  161, 164, 165, 173, 197, 223,
  336

topology, 36, 94, 138, 142, 143, 145,
  146, 148, 149, 151, 155, 158,
  188, 196, 203, 206, 255, 257,
  333

trajectory, 131, 264, 267, 289, 302

Transverse-electric, 36, 37

transverse-magnetic, 36, 40

tunability, 26, 29, 34, 41, 62, 63, 66,
  71, 75, 135, 234, 281

tunable-resonance, 27

Ultra-wideband, 211, 319, 335

UWB, 199, 211

vacuum, 33

varactor, 29, 30, 41, 93, 245, 332, 340

velocity, 62, 257

Virtualization, 320

wavefront, 30, 53, 55, 58, 61, 100,
  101, 105, 120, 231, 235, 237,
  274, 278, 309, 337

wideband, 323

wireless, 3, 79, 142, 161, 162, 173,
  177, 186, 208, 210, 211, 214,
  219, 221, 248, 253, 269, 277,
  282, 283, 286, 287, 294, 295,
  297, 300, 308, 309, 317, 319,
  320, 322, 329, 330, 336, 340

x-polarized, 67

X-Window, 219

y-polarized, 54, 72